General Preface to the Series

Because it is no longer possible for one textbook to cover the whole field of biology while remaining sufficiently up to date, the Institute of Biology has sponsored this series so that teachers and students can learn about significant developments. The enthusiastic acceptance of 'Studies in Biology' shows that the books are providing authoritative views of biological topics.

The features of the series include the attention given to methods, the selected list of books for further reading and, wherever possible, suggestions for practical work.

Readers' comments will be welcomed by the Education Officer of the Institute.

1980   Institute of Biology
                                                41 Queen's Gate
                                                London SW7 5HU

Preface

Autotrophic organisms synthesize biological matter from inorganic compounds. This involves the assimilation of carbon dioxide into sugars and the incorporation of inorganic nitrogen and sulphur into amino acids. The energy required for these processes is obtained from non-biological sources. Heterotrophic organisms on the other hand use organic compounds synthesized by autotrophs as their sources of carbon and energy. Many heterotrophs are also dependent on other organisms for a supply of amino acids as sources of nitrogen and sulphur.

This book compares the requirements of autotrophs and heterotrophs for carbon, nitrogen, sulphur and energy in relation to the synthesis of biological matter. It aims to demonstrate that the great diversity of energy-generating mechanisms and the synthesis of biological matter from organic and inorganic compounds by organisms is consistent with the principles of bioenergetics. These principles provide the framework for understanding the cycling of elements within ecosystems.

I would like to express my gratitude to Dr R. E. Williamson for many helpful comments in preparing this book.

Melbourne, 1980   J. W. A.

Contents

General Preface to the Series ... iii

Preface ... iii

1 **Transformations of Carbon, Nitrogen and Sulphur by Organisms** ... 1
1.1 Requirements of organisms for carbon and energy 1.2 Requirements of organisms for other elements 1.3 Bio-energetics of oxidizing and reducing various forms of carbon, nitrogen and sulphur 1.4 Coupling of endergonic and exergonic reactions in cells 1.5 Role of ATP and reducing agents as intracellular energy sources

2 **Primary Energy-generating Mechanisms of Autotrophs** ... 13
2.1 Green photosynthetic bacteria 2.2 Purple photosynthetic bacteria 2.3 Blue-green algae 2.4 Eukaryotic photoautotrophs 2.5 Energetics of photo-oxidation of photosynthetic pigments 2.6 Chemoautotrophic bacteria

3 **Assimilatory Processes of Autotrophic Organisms** ... 25
3.1 CO_2 assimilation in green plants 3.2 CO_2 assimilation in blue-green algae 3.3 CO_2 assimilation in anaerobic photosynthetic bacteria 3.4 CO_2 assimilation in chemoautotrophic bacteria 3.5 Some other assimilatory processes of green plants

4 **Energy Requirements of Heterotrophs** ... 35
4.1 The heterotrophic mode of life 4.2 Bioenergetics of the oxidation of organic carbon 4.3 Oxidation of organic carbon and synthesis of ATP by heterotrophs

5 **Interdependence of Organisms for Carbon, Nitrogen, Sulphur and Energy** ... 47
5.1 General principles 5.2 Aerobic ecosystems with green plants as the primary producers 5.3 The Galapagos Rift ecosystem 5.4 Aquatic ecosystems involving an anaerobic component

Further Reading ... 60

1 Transformations of Carbon, Nitrogen and Sulphur by Organisms

1.1 Requirements of organisms for carbon and energy

Living matter is comprised of a large number of organic molecules. The essential features of organisms such as their ability to grow, reproduce and respond to environmental stimuli depend on an ordered arrangement of organic molecules in cells and the maintenance and control of this arrangement. The element carbon forms the basic framework of the organic molecules found in organisms. Clearly the carbon in these molecules must be derived either directly or indirectly from the non-biological sources of carbon in the physical environment. The regions of the earth in which organisms grow is referred to as the biosphere. It includes the outer few metres of the earth's crust, the atmosphere and bodies of water such as oceans, lakes, ponds, rivers, etc. Neglecting the present day reserves of carbon of biological origin such as organisms themselves, leaf detritus, soil organic matter, etc., practically all of the carbon in the biosphere occurs as CO_2 (either in the air or in solution) and as various carbonates.

Organisms can be classified according to the sources of carbon that they use. Organisms which use the CO_2 of the physical environment are referred to as autotrophs (*auto*, Gk., self; *trophikos*, Gk., nourishment). They include green plants and various bacteria. Other organisms are unable to use CO_2. They require a supply of organic forms of carbon and are known as heterotrophs (*hetero*, Gk., other). They include animals, fungi and most (but not all) bacteria. It follows that heterotrophs must derive their organic carbon either directly or indirectly from autotrophs. After passage through a series of different heterotrophs, the organic carbon is eventually oxidized to CO_2 by the process of respiration (Fig. 1–1) thereby returning inorganic carbon to the physical environment. This is important since autotrophs would deplete the biosphere of the CO_2 that they need for their function within about 300 years. It is apparent therefore that autotrophs and heterotrophs are interdependent with respect to carbon.

Organisms must exist in a higher energy state than the environment to maintain order and control over their activities. It follows that both autotrophic and heterotrophic organisms have a fundamental requirement for energy. The energy requirements of autotrophs and heterotrophs are closely related to the carbon sources that they use. Autotrophs have a particularly high requirement for an external energy source for the reductive reactions associated with the synthesis of organic molecules (biological matter) from CO_2. This can be readily appreciated

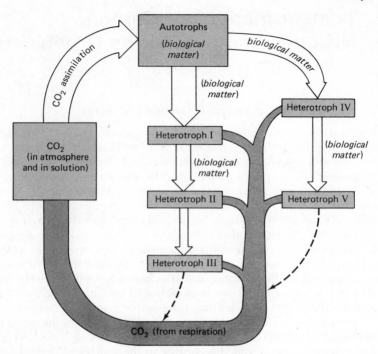

Fig. 1–1 Cycling of carbon between autotrophs and a series of heterotrophs. The width of the arrows provides a measure of the carbon flux for each component of the cycle. In this example heterotroph I could represent a herbivore, heterotroph II, a predator and heterotrophs III, IV and V, organisms which decompose biological matter (e.g. fungi and bacteria) following the deaths of the previous organisms in the cycle. In nature countless heterotrophs are involved in the cycle, each being dependent on other organisms in their own unique way. Collectively they would quantitatively oxidize the biological matter produced by autotrophs to CO_2 by respiration. The long term additions and losses of carbon in the cycle (e.g. weathering of rocks, deposition of biological matter in sediments) are not shown in the diagram.

by considering the reaction in the opposite direction. Most of us, for example, have enjoyed the energy released as heat during the combustion of wood (biological matter from plants) to CO_2. This illustrates in a subjective way how energy must have been gathered from the environment by plants and expended in synthesizing wood from CO_2. In green plants the energy required for the synthesis of biological matter from CO_2 is obtained from sunlight. Autotrophs which use light as their energy source are referred to as photoautotrophs. Some autotrophs derive their energy from the oxidation of reduced forms of various elements present in the biosphere. They are referred to as chemoautotrophs. It follows that environmental sources of energy are conserved in the biological matter synthesized by autotrophs and that this, when oxidized, is a potential source of energy for other cellular activities. In autotrophs this is especially important when the appropriate environmental source of energy (e.g. light) is temporarily unavailable (e.g. night-time in photoautotrophs). Heterotrophs do not require environmental sources of energy. They satisfy their energy requirements by oxidizing a portion of their organic carbon source in the process known as respiration. Heterotrophs exhibit considerable diversity in the compounds they use to oxidize organic matter. In addition to oxygen they include nitrate, sulphate and, in the absence of these compounds, certain carbon oxidation products produced by heterotrophs themselves (e.g. pyruvate).

Another matter related to the carbon and energy requirements of organisms is that the controlled synthesis of biological matter, which effectively conserves environmental sources of energy as chemical energy, must be isolated from the environment itself. This is achieved by the plasma membrane which surrounds all living cells and which is selectively permeable to particular compounds. In simple terms the plasma membrane of autotrophic organisms prevents the loss of high potential energy organic compounds from cells to the environment whilst permitting the inward passage of CO_2, inorganic ions and water. The plasma membrane of all organisms permits the inward passage of certain key organic metabolites. Clearly this is especially important in heterotrophs.

1.2 Requirements of organisms for other elements

In addition to carbon, oxygen and hydrogen which occur in almost all of the many organic compounds found in cells, organisms also require many other elements (Table 1). With the exception of carbon, nitrogen and sulphur these elements occur in biological matter in the same oxidation state as in the physical environment. In theory, all organisms could use environmental sources of these elements (e.g. PO_4^{3-}, Ca^{2+}, Cu^{2+}, K^+, etc.) and compete with each other to fulfil their requirements. In practice, many organisms obtain their supplies of these elements from other organisms; herbivores for example obtain most of their supplies of Ca^{2+} and Mg^{2+} from plant material. However, since all organisms require these elements in the same oxidation state and, with some qualifications, do not alter the oxidation state within their cells, it follows that the requirement of an organism for these elements is not dependent on the metabolism of the elements by previous organisms in the food chain.

The requirements of organisms for nitrogen and sulphur are of particular interest as these elements, unlike those described above, are subject to oxidation and reduction in a manner analogous to carbon. As for carbon, at least some organisms must be able to use the forms of

Table 1   Abundance and oxidation states of the major elements found in organisms. The oxidation states of the elemental forms available to organisms in the physical environment are also shown. The abundance of the elements in organisms is expressed as atoms of the element relative to 1000 atoms of carbon. Hydrogen and oxygen which are common to almost all of the organic compounds found in organisms and which are obtained principally from water are not shown in the table. Most of the elements required in only trace amounts are also not included.

	Elemental abundance		Elemental forms	
Element	Zea may (maize)	Homo sapiens (man)	Physical environment	Biological matter
Carbon	1000	1000	CO_2	$-HCOH-, -CH_2-$
Nitrogen	28.6	143	NO_3^-, N_2	$-NH_2$
Potassium	6.5	6.0	K^+	K^+
Calcium	1.6	25.0	Ca^{2+}	Ca^{2+}
Phosphorus	1.8	21.6	PO_4^{3-}	PO_4^{3-}
Magnesium	2.0	1.4	Mg^{2+}	Mg^{2+}
Sulphur	1.5	5.2	SO_4^{2-}	$-SH$
Chlorine	1.1	2.8	Cl^-	Cl^-
Iron	0.4	0.05	Fe^{2+}/Fe^{3+}	Fe^{2+}/Fe^{3+}

nitrogen and sulphur available in the physical environment. They include the free gaseous N_2 of the atmosphere and nitrate and sulphate in soils and bodies of water. In organisms, nitrogen is found in many classes of molecules, including the two most fundamental molecules of life, the nucleic acids and the proteins. Proteins also contain sulphur. The nitrogen found in these molecules occurs in low (reduced) oxidation states. This implies that those organisms which can use N_2 (gas), nitrate and sulphate as sources of nitrogen and sulphur must reduce these elements to the oxidation states found in the nucleic acids and proteins. These reactions, like the reduction of CO_2 to sugars, require energy. Green plants assimilate oxidized forms of nitrogen and sulphur into biological matter in reactions which use light as the environmental source of energy. It is therefore apparent that green plants are (and must be) completely self-sufficient in synthesizing biological matter from the resources of the physical environment. Other autotrophs use environmental sources of energy for incorporating inorganic forms of nitrogen and sulphur into organic matter although some of these are associated with environments in which nitrogen and sulphur occur in reduced forms (e.g. NH_3 and H_2S, see Table 2) and therefore do not need to expend energy in effecting their reduction. Many heterotrophs, including non-ruminant animals, are unable to reduce oxidized forms of

Table 2   Forms of carbon, nitrogen and sulphur used by some organisms.

Organism	Carbon	Nitrogen	Sulphur
Aerobic autotrophs			
Plants	CO_2	NO_3^-	SO_4^{2-}
Thiobacillus (bacterium)	CO_2	NO_3^-	$S_2O_3^{2-}, S^{2-}$
Anaerobic autotrophs			
Chlorobium (bacterium)	CO_2	N_2, NH_4^+	S^{2-}
Heterotrophs			
Penicillium (fungus)	Organic C	NO_3^-	SO_4^{2-}
Monogastric animals	Organic C	Organic N	Organic S

nitrogen and sulphur and are therefore dependent on other organisms (mostly autotrophs) for a supply of organic forms of nitrogen and sulphur as well as carbon. Some heterotrophs, however, are able to assimilate oxidized forms of nitrogen and sulphur into amino acids (Table 2). The energy required for these processes is derived from the oxidation of their organic carbon source.

The consumption of oxidized forms of nitrogen and sulphur from the environment by organisms would soon deplete the environment of available forms of these elements. Some organisms which are discussed later in this booklet catalyse the oxidation of reduced forms of nitrogen and sulphur, thus allowing these elements to recycle in much the same way as carbon.

1.3   Bioenergetics of oxidizing and reducing various forms of carbon, nitrogen and sulphur

The energy changes associated with the oxidation of various compounds can be quantified. Organic compounds such as glucose and fatty acids can be oxidized non-biologically with oxygen in a bomb calorimeter to determine the changes in free energy ($\Delta G'$) associated with the oxidations. Energy is released as heat in these processes implying that the reaction products (CO_2 and water) contain less free energy than the substrates. There has therefore been a loss of free energy (negative $\Delta G'$) and the reaction is said to be exergonic. Similarly NH_3 and H_2S (reduced forms of nitrogen and sulphur with low oxidation states as found in amino acids, etc.) can also be oxidized in exergonic reactions:

(1.1) $C_6H_{12}O_6 + 6O_2 \rightleftharpoons 6CO_2 + 6H_2O$ ($\Delta G' = -2870 \text{ kJ mol}^{-1}$)

(1.2) $C_{18}H_{34}O_2 + 25\frac{1}{2}O_2 \rightleftharpoons 18CO_2 + 17H_2O$ ($\Delta G' = -9800 \text{ kJ mol}^{-1}$)

(1.3) $NH_3 + 2O_2 \rightleftharpoons HNO_3 + H_2O$ ($\Delta G' = -715 \text{ kJ mol}^{-1}$)

(1.4) $H_2S + 2O_2 \rightleftharpoons H_2SO_4$ ($\Delta G' = -347 \text{ kJ mol}^{-1}$)

where $C_{16}H_{12}O_6$ and $C_{18}H_{34}O_2$ represent molecules of glucose and the fatty acid oleic acid respectively. The magnitude of the $\Delta G'$ values indicates how far the equilibrium in these reactions lies to the right since $\Delta G'$ (expressed in kJ mol^{-1}) is related to the equilibrium constant (K_{eq}) by the equation:

$$\Delta G' = -5.7 \log_{10} K_{eq} \text{ (see below for derivation)}$$

For the reaction:
(i)   Substrate(s) $\rightleftharpoons$ Product(s)

the free energy change (ΔG) is related to the heat content (ΔH) and change in entropy (ΔS) at temperature T by the expression

(ii) $\Delta G = \Delta H - T\Delta S$

For equation (ii) it can be shown that

(iii) $\Delta G = \Delta G° + RT \ln \dfrac{[\text{Product(s)}]}{[\text{Substrate(s)}]}$

where R is the gas content, T is temperature, $\Delta G°$ the standard free energy change when the products and substrates are present at unit concentration and ln $x = 2.303 \log_{10} x$. If we consider reaction (i) at equilibrium at 25°C then $\Delta G = 0$, $T = 298°K$ and [Product(s)]/[Substrate(s)] $= K_{eq}$ (i.e. equilibrium constant). Inserting these values and $R = 8.31$ J mol^{-1} °K^{-1} and 2.3 $\log_{10}$ for ln into equation (iii) then

$$0 = \Delta G° + 8.31 \times 298 \times \log_{10} K_{eq}$$

$$\Delta G° \simeq -5700 \log_{10} K_{eq} \qquad \text{(expressed in J mol}^{-1}\text{)}$$

or $\Delta G° \simeq -5.7 \log_{10} K_{eq}$ (expressed in kJ mol^{-1})

Since H$^+$ is a component of many biological reactions it follows that in these cases $\Delta G°$ applies at pH 0 (i.e. H$^+ = 1$ M). Accordingly the term $\Delta G'$ is used to refer to the standard free energy change at physiological pH.

Thus if $\Delta G'$ is -5.7 kJ mol^{-1}, K_{eq} is 10; if $\Delta G'$ is -11.4 kJ mol^{-1}, K_{eq} is 100. Conversely, if $\Delta G'$ is $+5.7$ kJ mol^{-1}, K_{eq} is 0.1; if $\Delta G'$ is $+11.4$ kJ mol^{-1}, K_{eq} is 0.01. It follows from this analysis that the substrates in reactions (1.1) to (1.4) are almost completely oxidized at chemical equilibrium. In cells, these reactions can serve as sources of energy. Most cells have energy-generating mechanisms based on the aerobic and/or anaerobic oxidation of glucose or related compounds (respiration) but some organisms described in Chapter 3 use the oxidation of NH$_3$ (and other forms of nitrogen with low oxidation states) as energy-generating mechanisms. The same principle applies to the oxidation of forms of sulphur with low oxidation states. Incidentally, reactions of this kind are important in a wider biological context as they participate in the recycling of nitrogen and sulphur in a manner analogous to that described for carbon.

Now consider reactions (1.1) to (1.4) in the opposite direction:

(1.5) $6CO_2 + 6H_2O \rightleftharpoons C_6H_{12}O_6 + 6O_2$ ($\Delta G' = +2870$ kJ mol^{-1})

(1.6) $18CO_2 + 17H_2O \rightleftharpoons C_{18}H_{34}O_2 + 25\frac{1}{2}O_2$ ($\Delta G' = +9800$ kJ mol^{-1})

(1.7) $HNO_3 + H_2O \rightleftharpoons NH_3 + 2O_2$ ($\Delta G' = +715$ kJ mol^{-1})

(1.8) $H_2SO_4 \rightleftharpoons H_2S + 2O_2$ ($\Delta G' = +347$ kJ mol^{-1})

The free energy changes of these reactions are positive and the reactions are said to be endergonic. It follows from the equation relating $\Delta G'$ and K_{eq} that it is not possible to produce significant amounts of products in reactions *(1.5)* to *(1.8)*. However, some organisms (e.g. plants) carry out the processes shown in reactions *(1.5)* to *(1.8)*. Clearly, organisms cannot carry out these reductions by the single step reactions as shown since good yields of product can only be obtained in reactions with a negative $\Delta G'$. The ability of organisms to carry out these seemingly impossible reactions lies in their capacity to couple various endergonic reactions (positive $\Delta G'$) with exergonic reactions (negative $\Delta G'$).

1.4   Coupling of endergonic and exergonic reactions in cells

Quantification of the free energy changes of reactions associated with the oxidation and reduction of carbon, nitrogen and sulphur facilitates our understanding of the synthesis of biological matter by organisms. However, before proceeding further, it is important to note that the velocity of a chemical reaction is unrelated to the free energy change. It should also be noted that exergonic reactions rarely proceed spontaneously. The combustion of wood to CO_2 for example is an exergonic reaction but requires a small input of energy (referred to as the activation energy) to initiate the reaction.

Let us consider the biological synthesis of compound B from compound A in an endergonic reaction:

(1.9)   A $\rightleftharpoons$ B   ($\Delta G' = +28.5$ kJ mol^{-1})

It follows from the equation relating $\Delta G'$ and K_{eq} that at chemical equilibrium there will be only one part of B to 100 000 parts of A. Now consider the exergonic reaction

(1.10)   X $\rightleftharpoons$ Y   ($\Delta G' = -34.2$ kJ mol^{-1})

At equilibrium there will be one million parts of Y to one part of X. The secret of a cell's ability to synthesize compounds like B in equation *(1.9)* is that they possess mechanisms for the coupling of endergonic and exergonic reactions. Cells contain thousands of different enzymes, each one of which is specific for a particular chemical reaction. They serve several important functions. Firstly they act as catalysts thus speeding up the rate of chemical reactions. Secondly they lower the activation energy of chemical reactions (Fig. 1–2) thus permitting them to proceed at physiological temperatures. However, enzymes in no way affect the $\Delta G'$ of a chemical reaction. But the important point that concerns us here is that some enzymes have the facility to couple exergonic and endergonic reactions. Let us imagine that a cell contains an enzyme which specifically couples reactions *(1.9)* and *(1.10)*. The enzyme has a molecular structure

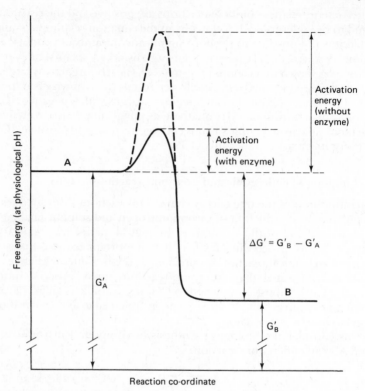

Fig. 1–2 Energy changes associated with a hypothetical exergonic reaction (A $\rightleftharpoons$ B) in the presence (———) and absence (–––––) of enzyme. Compounds A and B, at physiological pH, have inherent free energies denoted as G'_A and G'_B respectively. The free energy change ($\Delta G'$) is defined as the difference in free energy between product and substrate (i.e. $\Delta G' = G'_B - G'_A$). $\Delta G'$ is always negative for exergonic reactions. An enzyme decreases the activation energy for a given reaction but does not affect $\Delta G'$.

which determines that it will not catalyse separately either of the two component reactions. The reaction catalysed by the enzyme therefore is

(1.11) $A + X \longrightarrow B + Y$ $(\Delta G' = -5.7 \text{ kJ mol}^{-1})$

This reaction shows that a given number of molecules of A reacts with the same number of molecules of X and results in the production of the same number of molecules of B and Y. This is referred to as the stoichiometry of the reaction though it does not follow that all of the molecules of A and X which are present necessarily react together. The free energy change of reaction *(1.11)* is obtained by summing the $\Delta G'$ values of the two component reactions, i.e. $+28.5 + (-34.5) = -5.7$ kJ mol^{-1}. K_{eq} is therefore

10 and it follows that satisfactory yields of B can be attained at chemical equilibrium. This can be demonstrated by considering the case where 100 parts of A and 100 parts of B are supplied initially. It follows from the stoichiometry of equation *(1.11)* that equal parts of B and Y are synthetized at equilibrium. If the concentration of B and Y at equilibrium are denoted by n then the amount synthesized can be calculated as follows

$$K_{eq}=10=\frac{[B]\,[Y]}{[A]\,[X]}=\frac{n^2}{(100-n)^2}$$

Solving for n gives n = 76. Thus 76 parts of the original 100 parts of A have been metabolized to B. This represents a very different situation to that shown in equation *(1.9)*. The precise mechanisms by which enzymes achieve the catalysis of reactions such as that shown in equation *(1.11)* are exceedingly complex. Readers are referred to Study no. 42 in this series for more details.

1.5   Role of ATP and reducing agents as intracellular energy sources

Although autotrophs require large amounts of energy from the environment for the synthesis of biological matter from inorganic compounds, heterotrophs also require energy (though in lesser amounts) to support a large number of processes including active transport and the synthesis of nucleic acids and proteins from amino acids, processes common to all cellular organisms. Regardless of their mode of carbon nutrition and the energy generating mechanisms that they employ, organisms through their common ancestry, mediate their energy requirements via just a few key compounds. These compounds are either hydrolysed or oxidized in exergonic reactions and used as energy sources for coupling to endergonic reactions. The exergonic reactions can be classified into two groups.

1.5.1   Hydrolysis of high energy compounds

By far the most important reaction in this group is the hydrolysis of ATP. In cells, hydrolysis of ATP usually occurs in one of two ways:

(1.12)   ATP $\rightleftharpoons$ ADP + P$_i$ ($\Delta G'=-30.5$ kJ mol^{-1})

(1.13)   ATP $\rightleftharpoons$ AMP + PP$_i$ ($\Delta G'=-36.0$ kJ mol^{-1})

where P$_i$ and PP$_i$ represent orthophosphate and pyrophosphate respectively. Note that the $\Delta G'$ of these reactions is > -30 kJ mol^{-1}. This value is much more negative than the free energy of hydrolysis of most biological compounds which are typically less than -20 kJ mol^{-1}. For this reason ATP is known as a high energy compound. Although reaction *(1.13)* is no less important than reaction *(1.12)* in cells we shall for simplicity only consider reaction *(1.12)* in the examples that follow.

Another hydrolytic reaction with a highly negative $\Delta G'$ which occurs in cells is the hydrolysis of the compound acetyl coenzyme-A (acetyl CoA):

(1.14) Acetyl coenzyme-A $\rightleftharpoons$ Acetate + Coenzyme-A

$$(\Delta G' = -31.4 \text{ kJ mol}^{-1})$$

It is especially important in the synthesis of fatty acids and some amino acids though it is not relevant to the themes to be developed in this book. Other hydrolytic reactions with $\Delta G'$ values in excess of -30 kJ mol^{-1} include the following:

(1.15) Phosphopyruvate $\rightleftharpoons$ Pyruvate + P_i $(\Delta G' = -61.9 \text{ kJ mol}^{-1})$
(1.16) Diphosphoglycerate $\rightleftharpoons$ Phosphoglycerate + P_i

$$(\Delta G' = -49.4 \text{ kJ mol}^{-1})$$

Relative to the hydrolysis of ATP and acetyl coenzyme-A they are of minor importance as exergonic reactions for coupling to endergonic reactions in cells though they are important in the synthesis of ATP itself (see Chapter 4).

Although the free energy of hydrolysis of ATP is important in understanding how biological syntheses are possible this only begs the question how ATP is made from its hydrolytic products ADP and P_i, i.e.

(1.17) ADP + P_i $\rightleftharpoons$ ATP $(\Delta G' = +30.5 \text{ kJ mol}^{-1})$

Since the free energy of hydrolysis of ATP is used as an energy source by autotrophs for the synthesis of organic molecules from CO_2 it is apparent that autotrophs must use environmental sources of energy for the synthesis of ATP from ADP and P_i. Conversely the only sources of energy available to heterotrophs for the synthesis of ATP are derived from the oxidation of organic molecules in the processes of respiration. It is therefore apparent that the primary mechanisms of ATP synthesis in autotrophs and heterotrophs must represent important differences between the two groups of organisms.

1.5.2   Oxidation of reducing agents

The number of compounds which are oxidized in exergonic reactions and used in mediating energy transfer in organisms is similarly few. These reactions are no less important than those involving the hydrolysis of ATP etc.; indeed the free energy of oxidation of some compounds is used for the synthesis of ATP in most organisms. Some of the more important reducing agents found in cells, the structures of which are quite complex, are shown in Table 3. The redox potentials of these compounds at physiological pH (E_0'), together with those for oxygen and nitrate which serve as oxidizing agents in some organisms, are also shown. The free energy change associated with the oxidation of the biological reducing agents varies with the oxidizing agent and can be calculated from the differences in redox potential $(\Delta E_0')$ between the reductant and oxidizing agent. Consider the two component redox reactions (half-reactions):

Table 3   Some important biological reducing agents and oxidizing agents. All $\Delta G'$ values are shown for $2e^-$ transfers.

Compound	Oxidized form	Reduced form	E_0' (V)	$\Delta G'$ of oxidation of reduced form (kJ mol^{-1}) by oxygen	by nitrate
Biological reducing agents					
Ferredoxin (from spinach)	Fd$_{ox}$	Fd$_{red}$	−0.43	−239	−164
Nicotinamide adenine dinucleotide phosphate	NADP	NADPH$_2$	−0.32	−218	−143
Nicotinamide adenine dinucleotide	NAD	NADH$_2$	−0.32	−218	−143
Flavin adenine dinucleotide	FAD	FADH$_2$	−0.22	−199	−124
Oxidizing agents (from environment)					
Oxygen	O$_2$	H$_2$O	+0.81		
Nitrate	NO$_3^-$	NO$_2^-$	+0.42		

The redox potential of a half reaction (e.g. $\frac{1}{2}O_2 + 2H^+ + 2e^- \longrightarrow H_2O$) is measured as the potentiality of the system to furnish electrons to a hydrogen electrode at pH o (reference o V). Reduced forms of compounds listed in Table 3 can, in theory, reduce oxidized forms of compounds lower in the table in exergonic reactions.

(1.18) $A + 2e^- \longrightarrow B$ $(E_0' = +0.25$ V$)$

(1.19) $X + 2e^- \longrightarrow Y$ $(E_0' = -0.25$ V$)$

where A is the oxidizing agent and X is the reducing agent. The free energy change for the overall reaction

(1.20) $A + X \longrightarrow B + Y$

can be calculated from the expression

(1.21) $\Delta G' = -nF\Delta E_0'$

where n = number of electrons transferred from the reductant to the oxidizing agent and F = the Faraday constant (96.5 kJ mol^{-1}V^{-1}). For the reaction shown in equation (1.20) it is evident from the component reactions (1.18) and (1.19) that n = 2 and $\Delta E_0' = +0.25 - (-0.25) = 0.50$ V. Substituting into equation (1.21) then

$$\Delta G' = -2 \times 96.5 \times 0.5 = -96.5 \text{ kJ mol}^{-1}$$

Some idea of the potential free energy of the biological reducing agents can be demonstrated by calculating the free energy changes associated with their oxidation by oxygen. These values are shown in Table 3 though it should be noted that the biological reducing agents are not necessarily oxidized by oxygen but are used directly for energy-requiring processes.

The potential free energy of oxidation of $NADPH_2$, for example, is commonly employed to effect the reduction of oxidized forms of carbon in autotrophic organisms. In other organisms $NADH_2$ is oxidized by nitrate and the free energy change associated with this reaction (see Table 3) is used to effect the synthesis of ATP.

This analysis of the free energy changes associated with the oxidation of the important biological reducing agents aids our understanding of how the energy requirements of the endergonic reactions involved in the synthesis of biological matter in cells are fulfilled. However, this presents another more important problem, analogous to the regeneration of ATP from ADP and P_i. It concerns how the reduced forms of the biological reducing agents are regenerated from their oxidized forms, e.g.

$$(1.22) \qquad NADP + 2H^+ + 2e^- \longrightarrow NADPH_2$$

Here again it is evident that autotrophs must use environmental sources of energy and reducing power to effect these reductions whereas in heterotrophs they can only be achieved by oxidation of the organic carbon source. It is therefore evident that the different mechanisms for regenerating ATP and reducing agents in autotrophs and heterotrophs constitute important secondary characteristics of these two groups of organisms.

2 Primary Energy-generating Mechanisms of Autotrophs

It was established in the previous chapter that autotrophs are defined as organisms which use CO_2 as their sole carbon source for the synthesis of cellular material. It was also noted that environmental sources of energy are required for this process and that the energy requirements are supplied by ATP and biological reducing agents. This chapter is concerned with how ATP and reducing agents are generated from environmental sources of energy by photoautotrophs and chemoautotrophs. For photoautotrophs these processes are referred to as the light reactions of photosynthesisis. This nomenclature, however, cannot be applied to chemoautotrophs.

2.1 Green photosynthetic bacteria

The members of the bacterial family Chlorobacteriaceae (e.g. *Chlorobium*) are the simplest of the present-day photoautotrophs. They are green prokaryotic organisms which contain the pigment bacteriochlorophyll located within simple vesicles. In addition to CO_2 and light the Chlorobacteriaceae require anaerobic conditions and either hydrogen or a reduced form of sulphur such as H_2S for their growth. Since they are most commonly found in environments rich in H_2S (e.g. stagnant ponds and muds) they are usually referred to as the green photosynthetic sulphur bacteria.

The light-dependent incorporation of CO_2 by *Chlorobium* is accompanied by the stoichiometric oxidation of H_2S to elemental sulphur (S^0) according to the equation:

$$(2.1) \qquad 2H_2S + CO_2 \xrightarrow{h\upsilon} [CH_2O] + 2S^0 + H_2O$$

where $[CH_2O]$ refers to biological matter. This reaction implies that electrons from H_2S are used for the reduction of CO_2. However, H_2S is not a sufficiently strong reducing agent to effect the reduction of CO_2 (i.e. E_0' of H_2S is insufficiently negative). It must therefore be inferred that the requirement for light is to effect the elimination of electrons from H_2S (thereby causing its oxidation) which are trapped by a compound with a more negative redox potential. However H_2S itself is colourless and therefore does not absorb visible light to any appreciable extent. Light-dependent oxidation of H_2S must therefore be mediated via some pigment which absorbs visible light. In *Chlorobium* bacteriochlorophyll serves this function. It is found that certain of the bacteriochlorophyll molecules (designated P_{890}) are bleached when briefly illuminated with visible light. The bleaching is caused by the absorption of a photon of light, the energy of which can be readily calculated. The input of energy effects the photo-oxidation of the light acceptor molecules with the loss of an electron:

$$(2.2) \qquad P_{890} \xrightarrow{h\upsilon} P_{890}^+ + e^- \qquad (E'_0 = +0.47 \text{ V})$$

The ejected electron contains sufficient energy to cause the reduction of an electron acceptor X (i.e. $X + e^- \longrightarrow X^-$) which has an extremely negative redox potential. In *Chlorobium* the exact nature of X is uncertain but the E'_0 is known to be about -0.4 V. It follows from section 1.5.2 that reduced X (i.e. X^-) can in theory effect the reduction of the oxidized form of any compound with a more positive redox potential. In *Chlorobium* X^- is used to reduce oxidized ferredoxin (Fd_{ox}):

$$(2.3) \qquad X^- + Fd_{ox} \rightleftharpoons X + Fd_{red}$$

This reaction, which is weakly exergonic, serves to regenerate the primary electron acceptor X in addition to synthesizing the reduced form of a key biological reducing agent. Reduced ferredoxin can be used as a source of reducing equivalents (i.e. energy) for the synthesis of biological matter from CO_2. Furthermore, since reduced ferredoxin has the most negative redox potential of all biological reducing agents (except X^-) it can reduce other important biological reducing agents in exergonic reactions, e.g.

$$(2.4) \quad 2Fd_{red} + NAD + 2H^+_{ox} \longrightarrow 2Fd + NADH_2 \, (\varDelta G' = -21 \text{ kJ mol}^{-1})$$

However, this sequence of reactions is not complete unless oxidized bacteriochlorophyll (P_{890}^+) is returned to its original (unoxidized) state. Compounds with a redox potential more negative than bacterio-chlorophyll can effect this reaction. In *Chlorobium* H_2S and hydrogen serve as electron donors for the reduction of P_{890}^+:

$$(2.5) \qquad 2P_{890}^+ + H_2S \rightleftharpoons 2P_{890} + S^0 + 2H^+$$

$$(2.6) \qquad 2P_{890}^+ + H_2 \rightleftharpoons 2P_{890} + 2H^+$$

This explains the requirement for these substrates for the growth of *Chlorobium* and the products that they produce. The events shown in reactions (2.2) to (2.5) are summarized in Fig. 2–1. The free energy changes can be denoted by plotting the reaction sequence as a function of the redox potentials of the components; downward changes represent exergonic reactions and upward changes denote reactions which require an input of energy. This shows the fundamental importance of the photo-oxidation of bacteriochlorophyll.

Chlorobium, like all autotrophs, requires ATP as well as reducing equivalents for the synthesis of biological matter from CO_2. The mechanism of ATP synthesis in photosynthetic organisms is extremely complex but we need only concern ourselves with the energy requirements. In *Chlorobium* it seems that a portion of the reduced

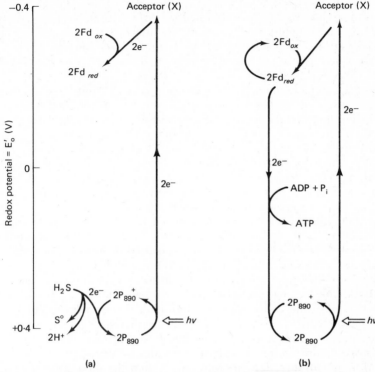

Fig. 2–1 Simplified schemes for (a) non-cyclic production of reduced ferredoxin using H_2S as electron donor and (b) cyclic photophosphorylation in anaerobic photosynthetic bacteria.

ferredoxin, generated in the presence of light, is used to reduce the oxidized form of bacteriochlorophyll (Fig. 2–1b). This reaction is highly exergonic and is more than sufficient to account for the energy required for the synthesis of ATP from ADP and P_i. This variant is known as cyclic photophosphorylation, so named because the electron cycles back to bacteriochlorophyll with the concomitant synthesis of ATP. One consequence of this process is that H_2S is not oxidized to elemental sulphur during cyclic photophosphorylation.

2.2 Purple photosynthetic bacteria

The purple photosynthetic bacteria, like the Chlorobacteriaceae, contain the pigment bacteriochlorophyll but typically appear purple or red due to the presence of carotenoid pigments which mask the bacterio-chlorophyll. The purple bacteria are divided into two families according to their anatomical and physiological properties.

The Thiorhodaceae or purple sulphur bacteria (e.g. *Chromatium*) are mobile organisms which exhibit a variety of morphological forms and contain bacteriochlorophyll structurally attached to membranes. The physiological characteristics of this family resemble those of the Chlorobacteriaceae in that they require light, anaerobic conditions and H_2S or hydrogen for the· assimilation of CO_2. Like the Chlorobacteriaceae they typically grow in anaerobic muds and stagnant waters exposed to light. The mechanisms for generating ATP and reducing equivalents are also similar although the sulphur produced from H_2S accumulates within the cells of the Thiorhodaceae whereas the Chlorobacteriaceae excrete sulphur into the growing medium.

The Athiorhodaceae (e.g. *Rhodopseudomonas* and *Rhodospirillum*) differ from the Thiorhodaceae in several respects. In the Athiorhodaceae bacteriochlorophyll is attached to membranes which are arranged in stacks. They can grow autotrophically in the presence of hydrogen, light and CO_2 but are unable to use reduced forms of sulphur from the environment as the electron donor for the reduction of oxidized bacteriochlorophyll *(2.6)*. They are therefore known as the purple non-sulphur bacteria. However, the Athiorhodaceae are generally regarded as organisms which use simple organic carbon compounds (e.g. acetate) rather than CO_2 as a source of carbon. Nevertheless, the assimilation of the simple organic compounds into biological matter in the organisms is enhanced by light. The Athiorhodaceae are therefore known as photoheterotrophic organisms and they demonstrate that not all photosynthetic organisms are necessarily autotrophic.

Although all three families of photosynthetic bacteria are able to use hydrogen as an electron donor for the reduction of oxidized bacteriochlorophyll (equation *(2.6)*), representatives of all three groups also exhibit light-dependent production of hydrogen. Net production of hydrogen is only observed when compounds such as H_2S (sulphur bacteria) or certain organic compounds (non sulphur bacteria) are used as electron donors. Furthermore, the production of hydrogen is inhibited by the presence of nitrogen which is reduced to ammonia (i.e. nitrogen fixation). The requirement for light for the evolution of hydrogen and reduction of nitrogen implies that light-generated reducing equivalents are diverted into H^+ and N_2 and that these two compounds compete for electrons:

(2.7)

$$6P_{890} \xrightarrow{h\upsilon} 6P_{890}^+$$

$$6H^+ \underbrace{\qquad}_{3H_2} 6e^- \underbrace{\qquad}_{2NH_3} N_2 + 6H^+$$

These processes do not appear to be related to a decrease in the rate of

CO_2 assimilation. It therefore seems as though photosynthetic bacteria use these reactions for discharging excess light-generated reducing equivalents under certain growth conditions.

2.3   Blue-green algae

The blue-green algae (Cyanophyceae) are also photosynthetic prokaryotes and for this reason are referred to as blue-green bacteria by some authorities. They nevertheless represent an important advance on the other photosynthetic prokaryotes. Although some blue-green algae grow under anaerobic conditions and a few can use H_2S as their electron donor, the majority are aerobic organisms and do not require a strong reducing agent such as H_2S or hydrogen for light-dependent incorporation of CO_2 into biological matter. Blue-green algae are therefore not restricted to anaerobic environments. In the light under anaerobic conditions, they reduce gaseous nitrogen to ammonia (nitrogen fixation) in a manner analogous to the other photosynthetic prokaryotes (equation (2.7)) though, as will be seen presently, different light-absorbing pigments are involved.

Under aerobic conditions blue-green algae exhibit light-dependent oxygen evolution at essentially the same rate as light-dependent CO_2 assimilation. By analogy with the light-dependent production of elemental sulphur from H_2S which accompanies the assimilation of CO_2 in the green and purple sulphur bacteria equation (2.1), the oxygen evolved by blue-green algae comes from water. This indeed is the case. Cells grown in solutions in which the oxygen atoms of water molecules contain the heavy isotope of oxygen (^{18}O) evolve gaseous $^{18}O_2$:

$$(2.8) \qquad 2H_2^{18}O + CO_2 \xrightarrow{h\upsilon} [CH_2O] + {}^{18}O_2 + H_2O$$

This implies that the reducing equivalents for the reduction of CO_2 are obtained from water. However, the redox potential for the reaction:

$$(2.9) \qquad H_2O \longrightarrow 2H^+ + \tfrac{1}{2}O_2 + 2e^-$$

is +0.81 V. This is far more positive than the redox potential of the P_{890} form of bacteriochlorophyll (+0.47 V) found in the other photosynthetic prokaryotes. Putting this another way, water, unlike H_2S or hydrogen, is not a sufficiently strong reducing agent to reduce oxidized bacterio-chlorophyll (P_{890}^+) in an exergonic reaction. It follows that the light-dependent production of a reducing agent with an E_0' sufficiently negative to effect the assimilation of CO_2 into organic matter in blue-green algae must entail the light-dependent oxidation of a pigment which has a redox potential more positive than water (+0.81 V). Furthermore, as will be discussed in section 2.6, photons of red light, which are known to be photosynthetically active in blue-green algae, impart barely sufficient energy to a pigment of redox potential +0.81 V to effect the reduction of

CO_2. On theoretical grounds, this suggests that more than one photo-oxidizable pigment system could be involved; two or more systems could be interconnected in such a way that electrons from water are progressively elevated in energy, first by one photo-oxidizable pigment and then by another to attain sufficient energy for the reduction of CO_2. By way of analogy, imagine two people (A and B) who want to place a small article on the top of a building during its construction. Both A and B can throw the article to 80% of the height of the building. However, if B is positioned on a platform halfway up the building, A can throw the article to B and B in turn can throw the article the remaining distance to the top of the building. In this analogy, the delivery of the article to the top is dependent on two stages; delivery is not achieved in the absence of either stage.

A phenomenon known as the 'red drop' provides an important lead to answering this problem. In blue-green algae the main light-absorbing pigments are chlorophyll-a which absorbs red and blue light, and the so called accessory pigment phycocyanin which mainly absorbs yellow-orange light. Chlorophyll-a and the cells of blue-green algae in which it is contained readily absorb red light of wavelengths 680–700 nm. However, when the effect of light of specific wavelengths on photosynthesis is examined it is found that per photon of light absorbed by the cells, light of 680–700 nm promotes relatively less photosynthesis than light of shorter wavelengths (i.e. light of 680–700 nm has a low photosynthetic efficiency). This shows that there is some feature of the light reaction of blue-green algae which prevents the utilization of the 680–700 nm light which it absorbs. However, it should be stressed that this applies only when 680–700 nm light is supplied in the absence of any other wavelengths. When the photosynthetic efficiency of 680–700 nm light is re-examined in the presence of a second (supplementary) beam of light (e.g. 620 nm) and corrections made for the photosynthesis that the supplementary beam supports, then it is found that the photosynthetic efficiency of 680–700 nm light increases to a value approximating that of shorter wavelengths. In other words there is some feature of the light-absorbing mechanism of blue-green algae which causes enhanced photosynthetic efficiency of 680–700 nm light when shorter wavelength light is also supplied. These data are consistent with the concept of two light reactions discussed earlier; one light-dependent reaction supplies a light-generated product which is required in the second light reaction, the product of the second reaction being required for the assimilation of CO_2. Various more complex experiments indicate that the first of the two light reactions involves the photo-oxidation of a pigment known as P_{680} which has a redox potential $>+0.85$ V. Light of wavelengths less than 680 nm, which is absorbed by the accessory pigment, is used in this reaction.

$$(2.10) \qquad P_{680} \xrightarrow{\ h\upsilon\ } P_{680}^{+} + e^{-} \qquad (E_0' > +0.85 \text{ V})$$

The second light reaction is excited by light of longer wavelengths (680–700 nm) absorbed by chlorophyll-*a*. This reaction involves photo-oxidation (bleaching) of a specialized chlorophyll-*a* molecule known as P_{700}:

$$(2.11) \qquad P_{700} \xrightarrow{\;h\upsilon\;} P_{700}^+ + e^- \qquad (E_0' = +0.45 \text{ V})$$

The important point here is that P_{700} has a redox potential similar to the P_{890} form of bacteriochlorophyll equation *(2.2)* whereas the redox potential of P_{680} is even more positive than the O_2/H_2O redox pair $(E_0' = +0.81$ V). By analogy with P_{890}, the photo-oxidation of P_{700} can effect the reduction of ferredoxin. However, the P_{700}^+ formed in this process cannot be returned to the P_{700} form by withdrawing an electron directly from water. The electron required for the reduction of P_{700}^+ is supplied by the other light reaction. When P_{680} is photo-oxidized the ejected electron is trapped by plastoquinone (PQ) thereby reducing it (PQH_2). The redox potential of PQ/PQH_2 (0.0 V) is considerably less than that of P_{700}/P_{700}^+ (Fig. 2–2). Reduced plastoquinone generated by the second light reaction can therefore reduce P_{700}^+ in a highly exergonic reaction. It is this reaction which links the two light reactions together thereby explaining the enhanced photosynthetic efficiency of 680–700 nm light when a supplementary beam of shorter wavelength is also supplied. However, one point remains. It is essential that P_{680}^+ generated by photo-oxidation of P_{680} is returned to its original state so that the chain of events can be repeated. This is achieved by withdrawing an electron from water. Here again the processes associated with the light reactions can be represented as changes in free energy by plotting the reactions as functions of the redox potentials of the component compounds (Fig. 2–2).

The chain of events shown in Fig. 2–2 is also relevant to the synthesis of ATP. Here it seems that the exergonic reaction associated with the reduction of $2P_{700}^+$ by reduced plastoquinone (PQH_2)

$$(2.12) \qquad PQH_2 + 2P_{700}^+ \longrightarrow PQ + 2P_{700} + 2H^+ \quad (\Delta G' = 86.9 \text{ kJ mol}^{-1})$$

is utilized as the energy source for the synthesis of ATP from ADP and P_i. Since this reaction forms part of an overall process in which electrons flow from water to ferredoxin with the concomitant phosphorylation of ADP, the process is known as non-cyclic photophosphorylation. The mechanism by which the synthesis of ATP is coupled to reaction *(2.12)* is quite complex (see HINKLE and McCARTY, 1978).

2.4   Eukaryotic photoautotrophs

The eukaryotic photoautotrophs include the true algae (green, brown and red), bryophytes, pteridophytes and higher plants. In these

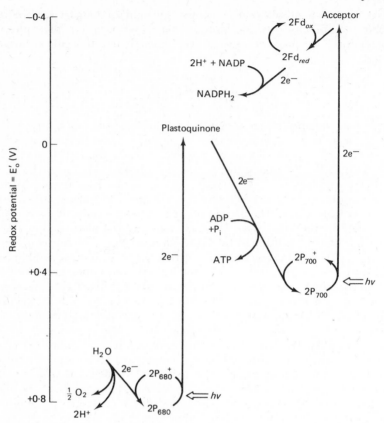

Fig. 2–2   Scheme for non-cyclic photophosphorylation in photosynthetic organisms which use water as electron donor.

organisms the light-dependent events are localized within chloroplasts. The mechanisms for the generation of reduced ferredoxin and ATP are essentially the same as in blue-green algae (Fig. 2–2) except that each group of organisms is characterized by different accessory pigments including various carotenoids, phycobilins and chlorophylls in addition to chlorophyll-a. The differences in the accessory pigments between the various groups presumably represent adaptations to the environments in which each group characteristically grows. Green algae for example are found close to the surface in a water profile and absorb red light. Red algae on the other hand grow deeper in the water profile where there is poor penetration of red light; they contain phycoerythrin which absorbs light of shorter wavelengths.

Although the photochemical events which lead to the production of reducing agents and ATP are localized in the chloroplasts of eukaryotic photoautotrophs, ATP and reducing equivalents are also required by the other subcellular organelles and compartments of the cell for their characteristic activities. The chloroplast is separated from the remainder of the cell by a double membrane referred to as the chloroplast envelope. This envelope is impermeable to both the reduced and oxidized forms of ferredoxin, NADP and NAD and is not freely permeable to ATP and ADP. The passage of reducing equivalents and high energy phosphate is facilitated by mechanisms known as shuttles. In summary they involve the reduction or phosphorylation of specific carrier molecules which are freely permeable to the chloroplast envelope. For example light-generated reducing equivalents in the form $NADPH_2$ are transferred to the carrier molecule oxaloacetate within the chloroplast, reducing it to malate. Malate passes through the chloroplast envelope to the cytoplasm where the reducing equivalents are transferred to NAD, thus oxidizing malate to oxaloacetate. The oxaloacetate returns to the chloroplast and recycles. In this example, the net effect is the transfer of light-generated reducing equivalents from $NADPH_2$ in the chloroplast to $NADH_2$ in the cytoplasm.

2.5   Energetics of photo-oxidation of photosynthetic pigments

So far we have assumed that a photon of photosynthetically active light provides sufficient energy to effect photo-oxidation of the light-absorbing pigment(s) of photosynthetic organisms with the simultaneous reduction of the appropriate electron acceptor. Let us take the case of the reduction of ferredoxin by photo-oxidation of P_{700} in organisms which use water as the electron donor:

$$(2.13) \qquad P_{700} + Fd_{ox} \xrightarrow{\ h\upsilon\ } P_{700}^+ + Fd_{red}$$

The free energy change of this reaction in the absence of light can be calculated from the equation $\Delta G' = -nF\Delta E_0'$. Thus

$$\Delta G' = -1 \times 96.5\ (-0.43-(+0.45)) = +84.9\ \text{kJ mol}^{-1}$$

We can calculate the energy supplied to this reaction by light. Let us consider red light of 670 nm since this is highly photosynthetically active and the intensity of this wavelength in sunlight is high. Einstein's law of photochemical equivalence predicts that a molecule of photosynthetic pigment reacts with exactly one photon of light energy. The energy (E) contained in a photon of wavelength λ and wave frequency of υ can be calculated from Planck's quantum theory:

$$E = h\upsilon$$

where $h =$ Planck's constant $(6.625 \times 10^{-34}\,\text{J s}^{-1})$ and $\upsilon = c/\lambda$ where c = speed of light $(3 \times 10^8\,\text{m s}^{-1})$.

Thus for one photon of light of 670 nm

$$E = \frac{hc}{\lambda} = \frac{6.625 \times 10^{-34} \times 3 \times 10^8}{670 \times 10^{-9}} = 2.966 \times 10^{-19} \text{J}$$

The energy absorbed by one gram mol of pigment when irradiated with light of 670 nm is obtained by multiplying by the Avogadro number ($N = 6.024 \times 10^{23}$).

Energy absorbed per mol $= NE = 6.024 \times 10^{23} \times 2.966 \times 10^{-19}$
$$= 17.867 \times 10^4 \text{J mol}^{-1}$$
or $= 179 \text{kJ mol}^{-1}$

It is therefore apparent that the energy absorbed from 670 nm light by a mol of P_{700} is more than sufficient to promote the reduction of ferredoxin by P_{700} ($\Delta G' = +85 \text{kJ mol}^{-1}$). Indeed it follows that the reaction in the presence of light has a negative free energy change. Similarly it can be shown that the light-dependent reduction of plastoquinone by P_{680} in green plants and the reduction of ferredoxin by P_{680} in photosynthetic bacteria also proceed with a negative free energy change.

Finally, we can calculate the theoretical quantum efficiency of photosynthetic organisms with two light reactions which use water as electron donor. The scheme in Fig. 2–2 predicts that two photons or quanta are required per electron removed from water. The assimilation of CO_2 into carbohydrate (denoted as $[CH_2O]$ in equation *(2.14)* requires 4 electrons:

It therefore follows that the theoretical number of quanta required for the assimilation of one molecule of CO_2 is eight. This is in general agreement with the values obtained by experiment. For red light of 670 nm the energy input is $8 \times 179 = 1432$ kJ mol^{-1} of CO_2 assimilated or $6 \times 1432 = 8592$ kJ mol^{-1} of glucose synthesized. The amount of energy conserved in glucose is obtained from the free energy of oxidation of glucose (-2870 kJ mol^{-1}) (see equation *(1.1)*). The conservation of absorbed light energy in glucose is therefore $2870/8592 \times 100 = 33\%$.

2.6   Chemoautotrophic bacteria

In certain localized parts of the physical environment, some elements are present in reduced forms (i.e. low oxidation state). These forms are typically associated with anaerobic environments such as stagnant ponds, muds and marshes, water-logged soils, etc. These conditions are also conducive to the formation of reduced forms of elements by biological processes (see Chapter 4). The stench associated with such environments is testimony to the presence of reduced forms of carbon, nitrogen and sulphur. However, even well-drained soils tend to contain a proportion of at least some biologically important elements in reduced oxidation states.

The exergonic reactions associated with the oxidation of reduced forms of elements other than carbon are exploited by chemoautotrophs as an energy source for the production of biological reducing agents and ATP. However, the conditions required by chemoautotrophs for growth are extremely critical as they require two classes of compound from the environment which tend to be incompatible. These involve both a fairly plentiful supply of an element in reduced form (e.g. NH_3, H_2S) and an oxidizing agent (e.g. oxygen) with a more positive redox potential than the reducing agent. It follows that chemoautotrophs are essentially aerobic organisms although they can grow at quite low partial pressures of oxygen and some can use oxidizing agents other than oxygen (e.g. nitrate). Being autotrophs they require CO_2 but unlike the photo-autotrophs they are colourless and do not require light. These requirements dictate the environmental conditions in which chemoautotrophs grow. They are commonly found near the surface of stagnant pools and other anaerobic bodies of water where there are supplies of reduced inorganic compounds, CO_2 and some oxygen.

Any one chemoautotroph exhibits some degree of specificity in the reduced inorganic compounds that it oxidizes. Collectively, however, they show considerable diversity in the reactions that they catalyse (Table 4). Nevertheless, the mechanisms for generating biological reducing agents and ATP in the various chemoautotrophs are essentially similar. They entail reduction of a c-type cytochrome by the inorganic reducing agent followed by subsequent electron flow via a series of biological redox carriers in exergonic reactions to oxygen (Fig. 2–3). The negative free energy change associated with the reactions is used for the synthesis of ATP. Alternatively reduced cytochrome-c can reduce NAD (they have similar redox potentials). It is therefore apparent that chemoautotrophs

Table 4 Some exergonic reactions used as primary energy sources by chemo-autotrophic bacteria.

Reaction	$\Delta G'$ (kJ mol^{-1})	Chemoautotrophs
$S^{2-} + 2O_2 \longrightarrow SO_4^{2-}$	-715	Thiobacillus
$S^0 + 1\frac{1}{2}O_2 + H_2O \longrightarrow SO_4^{2-} + 2H^+$	-502	Thiobacillus
$NH_3 + 1\frac{1}{2}O_2 \longrightarrow NO_2^- + H_2O + H^+$	-274	Nitrosomonas
$NO_2^- + \frac{1}{2}O_2 \longrightarrow NO_3^-$	-73	Nitrobacter
$H_2 + \frac{1}{2}O_2 \longrightarrow H_2O$	-234	Hydrogenomonas

exploit inherently unstable environmental conditions for the synthesis of ATP and the strong biological reducing agents for use in CO_2 assimilation.

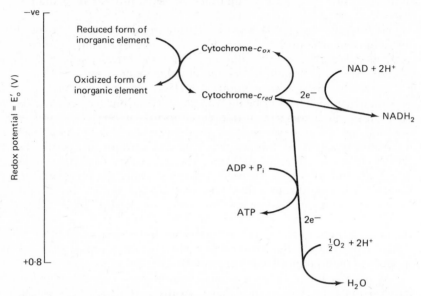

Fig. 2–3 Generalized scheme for the synthesis of reducing equivalents and ATP in chemoautotrophic bacteria. The scale of the ordinate varies with the E_0' of the inorganic redox pair.

3 Assimilatory Processes of Autotrophic Organisms

In Chapter 2 it was noted that autotrophic organisms utilize environmental sources of energy to effect the phosphorylation of ADP and the reduction of strong biological reducing agents. This chapter is concerned with the utilization of these energy sources for the synthesis of biological matter by autotrophs from the inorganic compounds present in the physical environment. It follows from the composition of biological matter (Table 1) that the reduction and assimilation of carbon from CO_2 must account for the greater part of the consumption of reducing agents and ATP in autotrophic organisms. However, the final section of this chapter demonstrates that the ATP and reducing equivalents produced by the light reaction of green plants can also be used for the assimilation of inorganic nitrogen and sulphur.

3.1 CO_2 assimilation in green plants

CO_2 assimilation in green plants is synonymous with the so-called dark reaction of photosynthesis. It involves the incorporation of CO_2 into hexoses in reactions which consume reducing equivalents and ATP but which do not in themselves require light. CO_2 assimilation takes place in the soluble matrix or stroma of chloroplasts which, in the light, is supplied with ATP and reducing equivalents from the light reactions. Details of the dark reaction and the associated bioenergetics can be found in HALL and RAO, 1977, and therefore only the more important points are summarized here. CO_2 (C_1) is incorporated one molecule at a time into the acceptor molecule ribulose diphosphate (C_5) to form 2 molecules of phosphoglycerate ($2 \times C_3$). The carboxylic acid groups (—COOH) in each of the 2 molecules of phosphoglycerate represent oxidized forms of carbon though in fact only one of these molecules contain the newly incorporated CO_2. These groups are phosphorylated by ATP, thereby 'exciting' them to form diphosphoglycerate before being reduced by $NADPH_2$ to form phosphoglyceraldehyde (Fig. 3–1). Since 2 molecules of phosphoglycerate are formed per molecule of CO_2 incorporated it follows that 2ATP and 2$NADPH_2$ are required per molecule of CO_2. The oxidation state of the aldehyde group (—CHO) of phosphoglyceraldehyde is the same as that found in the carbohydrates. Given a large number of molecules of phosphoglyceraldehyde, various rearrangements can occur to form various sugars containing 4, 5, 6, 7 and 8 carbon atoms. Indeed these reactions are used to regenerate the CO_2 acceptor molecule ribulose diphosphate while at the same time allowing a net accumulation of hexose sugar. An example is shown in Fig.

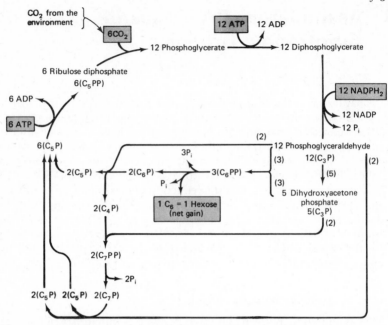

Fig. 3–1 An example of the stoichiometry of CO_2 assimilation showing the incorporation of 6 molecules of CO_2 with the net synthesis of 1 molecule of hexose and resynthesis of 6 molecules of the ribulose diphosphate. Compounds represented as C_4P and C_7PP denote phosphorylated sugars containing four and seven carbon atoms with one and two phosphate groups respectively.

3–1; it demonstrates that in the presence of 6 molecules of ribulose diphosphate 12 molecules of $NADPH_2$ and 18 molecules of ATP are required for the incorporation of 6 molecules of CO_2 with the net synthesis of a molecule of hexose (C_6) sugar and the regeneration of the 6 molecules of ribulose diphosphate used to initiate CO_2 assimilation. This scheme would be accompanied by the concomitant evolution of 6 molecules of oxygen by the light reaction.

On examining the scheme in Fig. 3–1 it is apparent that although ATP is used to form various phosphorylated intermediates in the cycle, the net products formed from 18 ATP during the synthesis of a molecule of hexose are 18 ADP and 18 P_1. The free energy change associated with this process is readily calculated from the data and principles discussed in Chapter 1. Similarly the free energy change associated with the oxidation of $NADPH_2$ can be readily calculated from the redox potentials of $NADP/NADPH_2$ ($E_0' = -0.32$ V) and O_2/H_2O ($E_0' = +0.81$ V) using the equation $\Delta G = -nF\Delta E_0'$. It might at first seem curious to use the redox potential of O_2/H_2O rather than CO_2/hexose in the calculation that follows. This is because the NADP produced during CO_2 assimilation in the reaction

(3.1) 12NADPH$_2$ + 6CO$_2$ + 18ATP $\longrightarrow$ Hexose + 6H$_2$O +
12NADP + 18ADP + 18P$_i$

is reduced back to NADPH$_2$ in a reaction using water as the eventual electron donor (see Chapter 2). Thus the total free energy change associated with the oxidation of NADPH$_2$ and hydrolysis of ATP during the synthesis of a hexose from 6 molecules of CO$_2$ can be calculated as follows:

18 × (ATP $\longrightarrow$ ADP + P$_i$)   ($\Delta G' = 18 \times -30.5 = -549$ kJ mol^{-1})
12 × (NADPH$_2$ + $\frac{1}{2}$O$_2$ $\longrightarrow$ NADP + H$_2$O)
($\Delta G' = 12 \times -218$ = $\underline{-2616}$ kJ mol^{-1})
-3165 kJ mol^{-1}

On the other hand the $\Delta G'$ for the oxidation of a hexose (e.g. glucose) to 6 molecules of CO$_2$ and water in a bomb calorimeter is -2870 kJ mol^{-1}. It is therefore apparent that the net free energy change for the biological synthesis of a hexose from 6 molecules of CO$_2 = +2870 + (-3165) = -295$ kJ mol^{-1}, i.e. the overall reaction proceeds with a negative $\Delta G'$. Thus at chemical equilibrium very significant yields of hexose will occur.

One other feature of CO$_2$ assimilation in green plants is also of considerable interest. The chloroplast in which both the light-dependent and light-independent reactions occur contains very low concentrations of NADP. In the light, isolated chloroplasts rapidly reduce their small internal supply of NADP by light-dependent electron flow from water (Fig. 2–2). This would be accompanied by the evolution of a small amount of oxygen (almost too small to measure) which would cease when all the NADP is reduced. However, if some suitable substrate which can be reduced by NADPH$_2$ is supplied, this causes the oxidation of NADPH$_2$ to NADP thus promoting further light-dependent reduction of NADP with the production of more oxygen. If relatively large amounts of substrate are supplied then the NADPH$_2$ is oxidized continuously thus causing the continuous evolution of oxygen at rates which can be determined readily. It is apparent from the pathway of CO$_2$ assimilation in leaves (Fig. 3–1) that oxidation of NADPH$_2$ is achieved by diphosphoglycerate. It therefore follows that isolated chloroplasts in the light evolve oxygen when supplied with diphosphoglycerate according to the following sequence:

This sequence shows that the light and dark reactions are coupled together and are therefore interdependent, i.e. oxygen evolution depends on the continuous oxidation of NADPH$_2$ by diphosphoglycerate. Furthermore, since illuminated chloroplasts also synthesize ATP it follows that the precursors of diphosphoglycerate shown in Fig. 3–1 (e.g. phosphoglycerate, ribulose diphosphate + CO$_2$) also induce oxygen evolution by isolated chloroplasts. Thus the light-independent or 'dark' reactions of the CO$_2$ assimilation pathway can be studied by measuring the oxygen evolved by the associated dark reaction. Illuminated chloroplasts also catalyse oxygen evolution when supplied with certain other biological substrates unrelated to the CO$_2$ assimilation pathway (e.g. nitrite and oxaloacetate). This implies that other metabolic processes also utilize reducing equivalents (and ATP) produced by the light reactions (see section 3.5).

The mechanism of CO$_2$ assimilation shown in Fig. 3–1 is common to all green plants and is known as the C$_3$ mechanism since phosphoglycerate, the first product of CO$_2$ incorporation, contains three carbon atoms. However, some plants, which are typically indigenous to hot dry climates have an ancillary mechanism for assimilating CO$_2$. This process, which is known as the C$_4$ mechanism, entails incorporation of CO$_2$ into the acceptor molecule phosphopyruvate to form oxaloacetate, a compound with four carbon atoms. However, the CO$_2$ is subsequently removed from a simple derivative of oxaloacetate and re-assimilated by the C$_3$ mechanism shown in Fig. 3–1. Plants with the C$_4$ mechanism (C$_4$ plants) lose less water per CO$_2$ molecule incorporated than plants which lack the C$_4$ mechanism (C$_3$ plants). C$_4$ plants are therefore better adapted to growth under hot dry conditions. They consume more ATP per molecule of CO$_2$ assimilated than C$_3$ plants (i.e. the energy requirements are greater than those shown in equation *(3.1)*) but this is unlikely to be a disadvantage given the high light intensities associated with hot dry environments.

3.2   CO$_2$ assimilation in blue-green algae

The mechanism of CO$_2$ assimilation in blue-green algae, is essentially the same as in green plants. The only important difference is that CO$_2$ assimiliation in blue-green algae occurs throughout the prokaryotic cell whereas in green plants it takes place in the stroma of chloroplasts.

3.3   CO$_2$ assimilation in anaerobic photosynthetic bacteria

The mechanism of CO$_2$ assimilation in most anaerobic photosynthetic bacteria is also similar to the C$_3$ mechanism of green plants except that NADH$_2$ rather than NADPH$_2$ serves as the reducing agent for the reduction of diphosphoglycerate. However, since anaerobic photosynthetic bacteria have only one light-reaction and use compounds such as H_2S and hydrogen as electron donors for the light reactions, the overall bioenergetics of CO_2 assimilation are quantitatively different to those of green plants although similar bioenergetic principles apply. This difference stems from the lower redox potentials for S^0/H_2S (E_0'=approx.-0.2 V) and H^+/H_2 ($E_0'=-0.42$ V) compared with O_2/H_2O ($E_0'=+0.81$ V).

In some photosynthetic bacteria (e.g. *Chlorobium thiosulfatophilum*) the activity of some of the enzymes of the C_3 pathway of CO_2 assimilation is insufficient to account for the rate of CO_2 assimilation by the bacterium. Further, when CO_2 labelled with the radioactive isotope ^{14}C is supplied to the bacterium in the light, ^{14}C-label is primarily associated with glutamate rather than phosphoglycerate as predicted for the C_3 mechanism in Fig. 3–1. This suggests that other processes contribute to CO_2 assimilation in these organisms. They contain two unusual enzymes which catalyse the following reductive carboxylation reactions:

(3.3)   Acetyl-CoA + CO_2 + 2 Fd$_{red}$ ⟶ Pyruvate + CoA + 2 Fd$_{ox}$

(3.4)   Succinyl-CoA + CO_2 + 2 Fd$_{red}$ ⟶ α-Oxoglutarate + CoA + 2 Fd$_{ox}$

These reactions are analogous to two essentially irreversible reactions of the TCA cycle of aerobic oxidation:

(3.5)   Pyruvate + CoA + NAD ⟶ Acetyl-CoA + CO_2 + NADH$_2$

(3.6)   α-Oxoglutarate + CoA + NAD ⟶ Succinyl-CoA + CO_2 + NADH$_2$

The important point here is that the enzymes of the TCA cycle are specific to NAD and those of the photosynthetic bacteria to ferredoxin. This confers important differences on the energetics of the two pairs of reactions since the redox potential of ferredoxin is substantially more negative than that of NAD. Putting this another way, reduced ferredoxin is a sufficiently powerful reducing agent to effect the reductive carboxylation of acetyl-CoA and succinyl-CoA whereas NADH$_2$ is not. The reduced ferredoxin required for reactions *(3.3)* and *(3.4)* is supplied by the light reactions. When $^{14}CO_2$ is supplied to *Chlorobium*, ^{14}C-label is initially incorporated into intermediates of the TCA cycle and their simple derivatives (e.g. α-oxoglutarate to glutamate, oxaloacetate to aspartate). It has been proposed therefore, that CO_2 incorporation in some photosynthetic bacteria proceeds, at least in part, via the reaction sequence shown in Fig. 3–2. In summary it involves a modified TCA cycle (see Chapter 4) running in reverse. The important modifications include the use of reduced ferredoxin as electron donor (equations *(3.3)* and *(3.4)*) and the involvement of several enzymes which catalyse the synthesis of oxaloacetate from acetate. Collectively these modifications alter the bioenergetics in such a way that the TCA cycle can run in reverse in the presence of light-generated reductant and ATP and so incorporate four molecules of CO_2 for each turn of the cycle.

3.4   CO$_2$ assimilation in chemoautotrophic bacteria

Carbon dioxide assimilation has been studied in several chemoautotrophs including *Thiobacillus ferrooxidans* (Fe^{2+} oxidizing), *T. thiooxidans* (sulphide oxidizing), *Nitrosocystis oceanus* (NH$_3$ oxidizing), *Nitrobacter agilis* (nitrite oxidizing) and several others. Although the principles for generating reducing equivalents and ATP in these organisms are quite different to the photoautotrophs, CO$_2$ assimilation in chemoautotrophs proceeds via the C$_3$ pathway as described for green plants and most anaerobic photosynthetic bacteria (Fig. 3–1). For example, in the presence of nitrite, most of the ^{14}CO$_2$ incorporated by cells of *Nitrobacter agilis* is initially present in phosphoglycerate. Thereafter the proportion of ^{14}C-label associated with phosphoglycerate declines and the proportion in sugar phosphates and sucrose is greatest after longer periods of incubation as predicted from Fig. 3–2.

When cells of *Nitrobacter agilis* are deprived of nitrite they exhibit very little CO$_2$ assimilation. The data in Table 5 demonstrate that extracts of

Table 5   Effect of various compounds on the incorporation of ^{14}CO$_2$ into biological matter by cell-free extracts of the chemoautotroph *Nitrobacter agilis*. (Data abstracted from Aleem, M. I. H. (1965). *Biochim. Biophys. Acta*, **107**, 14–28.)

Additions	^{14}C-label incorporated (counts s^{-1})	Additions	^{14}C-label incorporated (counts s^{-1})
Nil	50	NO$_2^-$	93
ATP or ADP	<67	NO$_2^-$ + ADP	172
NADH$_2$ or NADPH$_2$	<67	NO$_2^-$ + NAD	170
ATP + NADH$_2$	1433	NO$_2^-$ + ADP + NAD	605
ATP + NADPH$_2$	100	NO$_2^-$ + ADP + NADP	208

N. agilis catalysed CO$_2$ incorporation in the presence of NADH$_2$ and ATP but relatively insignificant amounts were incorporated when either substrate was omitted and when NADH$_2$ was replaced with NADPH$_2$. Extracts also failed to catalyse significant CO$_2$ incorporation when ATP was replaced with ADP and NADH$_2$ with NAD. However, when extracts were incubated with nitrite, ADP and NAD, significant CO$_2$ assimilation occurred implying that nitrite was serving as an energy source for the synthesis of ATP and NADH$_2$ for CO$_2$ assimilation. It is interesting to note that the chemoautotrophic bacteria, like the photosynthetic bacteria, use NADH$_2$ and not NADPH$_2$ as the reducing agent for CO$_2$ assimilation.

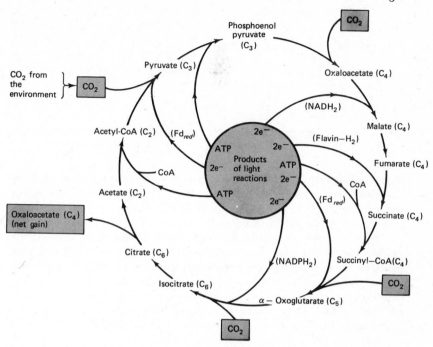

Fig. 3-2 The reductive carboxylic acid cycle of CO_2 assimilation as proposed for *Chlorobium thiosulfatophilum* and some other anaerobic photosynthetic bacteria. (Scheme adapted from Evans, M. C. W. *et al.* (1966). *Proc. Natl. Acad. Sci. U.S.*, **55**, 928–34.)

3.5 Some other assimilatory processes of green plants

3.5.1 Assimilation of inorganic nitrate

Plants utilize nitrate as their nitrogen source for the synthesis of the amino acids glutamine and glutamate which in turn serve as sources of nitrogen for the synthesis of most other nitrogen-containing compounds in plants. The incorporation of inorganic nitrate into glutamine and glutamate entails the reduction of nitrate by way of nitrite to ammonia. It is now recognized that many plants incorporate a proportion of their nitrate supply into glutamine and glutamate in photosynthetic cells. Within the photosynthetic cell the reduction of nitrite to ammonia and its incorporation into glutamine and glutamate occurs in chloroplasts and utilizes ATP and reduced ferredoxin from the light reactions as the source of energy (Fig. 3–3). Coupling of the reactions of nitrogen assimilation to the light reactions can be demonstrated by application of the principles described for carbon (e.g. equation *(3.2)*). For example, illuminated chloroplasts reduce nitrite to ammonia with the evolution of 1.5 molecules of oxygen per molecule of nitrite reduced as follows:

(3.7)

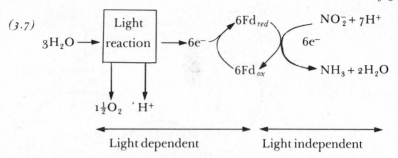

Light dependent   Light independent

Since light is required for the synthesis of reduced ferredoxin it follows that nitrite reduction in the dark is negligible. Light is also required for the synthesis of ATP consumed in reaction III (Fig. 3–3) and reduced ferredoxin consumed in reaction IV. Collectively, reactions II–IV (Fig. 3–3) can be considered as dark reactions of nitrogen assimilation analogous to the dark reactions of CO_2 assimilation. Chloroplasts, however, do not catalyse the reduction of nitrate, the form of inorganic nitrogen obtained from the environment. Reduction of nitrate to nitrite (reaction I) occurs in the cytoplasm. However, it is likely that at least some of the reducing equivalents required for nitrate reduction in photosynthetic cells are derived indirectly from the light reactions of chloroplasts. The membrane of the chloroplast is impermeable to $NADPH_2$ and ferredoxin but various small molecules serve as carriers to transport the reducing equivalents into the cytoplasm (Fig. 3–3).

3.5.2   Sulphate assimilation

The reduction and incorporation of inorganic sulphate into the sulphur amino acid cysteine affords another example of a reaction which utilizes ATP and reducing equivalents from the light reactions of the chloroplast. Chloroplasts catalyse the reduction of sulphate by way of sulphite (though sulphite remains attached to a carrier molecule) to sulphide. This overall reaction entails the consumption of one molecule of ATP and requires $8e^-$ supplied from reduced ferredoxin. The sulphide produced in this process is incorporated into an acceptor molecule (O-acetylserine) to form cysteine. Although it is uncertain whether externally added inorganic sulphite is metabolized in exactly the same manner as the sulphite formed from sulphate (which is attached to a carrier molecule), the data in Fig. 3–4 nevertheless show that the reduction of inorganic sulphite to sulphide by isolated chloroplasts is dependent on the light reactions. Figure 3–4 also shows that sulphide, produced in the light from sulphite, is rapidly incorporated into cysteine by chloroplasts on addition of O-acetylserine.

3.5.3   Other processes

Chloroplasts utilize light-generated ATP and reducing equivalents for

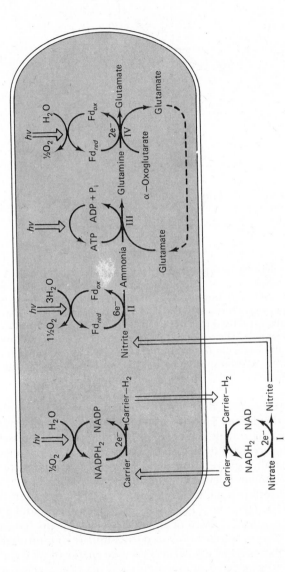

Fig. 3–3 Scheme for the light-dependent assimilation of inorganic nitrate into the amino acid glutamate in a photosynthetic cell of a green plant. The area enclosed with the double lines represents a chloroplast and the area outside denotes cytoplasm. Nitrate is reduced to nitrite in the cytoplasm by reducing equivalents supplied from the chloroplast via a carrier molecule which is transported across the chloroplast membrane. Nitrite is reduced to ammonia in the chloroplast and incorporated via glutamine into glutamate. The requirements for reduced ferredoxin and ATP are supplied by the light reactions with the concomitant evolution of oxygen.

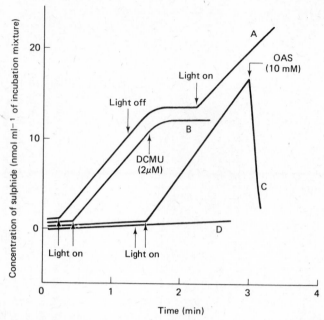

Fig. 3–4 Light-dependent reduction of inorganic sulphite to sulphide by isolated chloroplasts. Suspensions of pea chloroplasts (A to C) were supplied in the dark with 1 mM Na_2SO_3 and illuminated at the times shown. Experiment D was a control lacking sulphite. Other treatments were made as shown. The DCMU (dichlorophenyl dimethylurea) used in experiment B is a potent inhibitor of the light reactions of chloroplasts. The consumption of sulphide following addition of the sulphide acceptor molecule O-acetylserine (OAS) to experiment C was accompanied by the synthesis of an approximately equimolar amount of cysteine. (From Ng, B.H. and Anderson, J. W. (1979). *Phytochemistry*, **18**, 573–80.)

many processes in addition to the reductive assimilation of inorganic carbon, nitrogen and sulphur. They include the synthesis of fatty acids and incorporation of amino acids into proteins. Although these processes do not fall within the accepted definition of assimilation reactions they nevertheless show that the light reactions supply the energy requirements for virtually all the synthetic reactions of chloroplasts.

In summary, autotrophic organisms tap a diversity of energy sources (light, oxidation of H_2S, NH_3, etc.). The energy is trapped as ATP and, in different organisms, as one or more of several reducing agents (Fd_{red}, $NADPH_2$, $NADH_2$). These molecules are used to assimilate CO_2 through one of a number of pathways which harness the free energy of ATP and the reducing agents to give the overall assimilation reaction a negative $\Delta G'$. Additionally, they are used in numerous other energy-requiring syntheses such as the assimilation of inorganic nitrogen and sulphur into amino acids and the synthesis of fats, protein and nucleic acids.

4 Energy Requirements of Heterotrophs

4.1 The heterotrophic mode of life

As noted in Chapter 1, heterotrophs are defined as organisms which require an organic carbon source to satisfy their requirements for carbon and energy. In nature, dead or living biological matter invariably serves as the source of organic carbon for heterotrophs. With some exceptions, any carbon associated with large and complex molecules, is usually absorbed into the cells of heterotrophs as relatively simple molecules. Wood-rotting fungi for example, which use cellulose as their principal carbon source, secrete enzymes into the wood which hydrolyse cellulose to glucose. Thus, although cellulose serves as the carbon source, carbon is actually absorbed as glucose. Similarly monogastric animals degrade large biological molecules such as starch, glycogen and protein to their component monomeric units (sugars and amino acids) in their gastro-intestinal tract prior to absorption. These relatively simple organic molecules are used by heterotrophs for the synthesis of various complex biological molecules such as nucleic acids and proteins. It follows that heterotrophs require energy to effect the exact and ordered synthesis of these essential biological molecules. In addition, heterotrophs, like all organisms, require energy to create and maintain order and control over their activities. In many heterotrophs energy is required for movement (e.g. animals) and rapid growth (e.g. some fungi), characteristics which amongst other things enable heterotrophs to explore the environment for organic carbon sources.

Relative to autotrophs, heterotrophs require relatively little energy in the form of reducing equivalents since their oganic carbon source is already in a reduced form. A large proportion of the energy requirements of heterotrophs is derived from the free energy of hydrolysis of ATP. However, the production of reducing equivalents in the form of $NAD(P)H_2$ is an important function of most heterotrophs. While some $NAD(P)H_2$ is used in reductive reactions (e.g. the synthesis of fats) most of it is oxidized by an oxidizing agent which is usually derived from the environment (e.g. oxygen or nitrate). The oxidation of $NAD(P)H_2$ by oxidizing agents from the environment proceeds with a highly negative free energy change (Table 3). In organisms, oxidation of $NAD(P)H_2$ in this way is associated with the phosphorylation of ADP although it is important to note that many anaerobic organisms are unable to use environmental oxidizing agents. In these organisms ATP is synthesized by other mechanisms. The oxidation of organic carbon, regardless of the mechanism involved, is referred to as respiration. In heterotrophs the proportion of the organic carbon source respired varies greatly but is typically about 90%; the remaining 10% is built into new biological matter. However, autotrophs also exhibit respiration. Germinating seeds, for example, are heterotrophic; they have no mechanism for trapping an environmental source of energy (light) until they emerge and commence photosynthesis. In mature plants, some cells (e.g. root cells) are heterotrophic and all cells including photosynthetic cells exhibit respiration; this is especially important for maintaining the photosynthetic cell at night.

Heterotrophs exhibit considerable variation in the compounds they use as electron acceptors for the oxidation of organic molecules. The most common of these is molecular oxygen (which is then reduced to water) whereas some anaerobic organisms use the oxidized carbon product itself, e.g. reduction of pyruvic acid (produced by partial oxidation of glucose) to lactate or ethanol plus CO_2. Other anaerobic organisms have evolved mechanisms which enable them to utilize compounds from the environment such as nitrate, sulphate, ferric iron and H^+ as electron acceptors thereby producing the products NH_3 and N_2, H_2S and S^0, Fe^{2+} and H_2 respectively. Organisms of the latter type are well suited to growth in anaerobic environments containing relatively high concentrations of the appropriate electron acceptor. It is apparent that the reduced products formed by these organisms are reaction products resulting from the relevant respiratory pathways. These processes are therefore known as respiratory (or dissimilatory) nitrate reduction, respiratory (or dissimilatory) sulphate reduction, etc., and are quite unrelated to assimilatory nitrate and sulphate reduction described in Chapter 3. It is likely that a very small proportion of reduced inorganic nitrogen (NH_3) and sulphur (H_2S) is incorporated into organic molecules by organisms of this type but most of it is excreted into the growing medium where it accumulates often to quite high concentrations. In due course the reduced forms of nitrogen, sulphur, iron, etc., produced by dissimilatory heterotrophs can, in the presence of oxygen, be oxidized by chemoautotrophs and used as an energy source for CO_2 incorporation thereby establishing cycles for the circulation of nitrogen, sulphur, etc., in addition to carbon (see Chapter 5).

4.2 Bioenergetics of the oxidation of organic carbon

The overall free energy changes associated with some reactions involving the oxidation of glucose (which can be considered as a model source of organic carbon) are shown in Table 6. It is evident that the free energy changes for the partial anaerobic oxidation of glucose to lactate or ethanol plus CO_2 are relatively small compared with the complete oxidation of glucose to CO_2 by molecular oxygen. In organisms this difference is reflected in the amount of ATP synthesized. Aerobic organisms typically synthesize 36 molecules of ATP per molecule of

Table 6   Free energy changes associated with the oxidation of glucose by various organisms.

Organism	Oxidation reaction	$\Delta G'$ (kJ mol^{-1})	ATP molecules synthesized per molecule glucose
Aerobic			
Animals	Glucose + 6O$_2$ $\longrightarrow$ 6CO$_2$ + 6H$_2$O	-2870	36
Fermentative anaerobic			
Lactobacillus	Glucose $\longrightarrow$ 2 Lactate	-197	2
Yeast	Glucose $\longrightarrow$ 2 Ethanol + 2CO$_2$	-167	2
Dissimilatory anaerobic			
Parococcus denitrificans	Glucose + 4.8HNO$_3$ $\longrightarrow$ 6CO$_2$ + 2.4 N$_2$ + 8.4 H$_2$O	>-1945	$\simeq 20$
Desulfovibrio	Glucose + 3H$_2$SO$_4$ $\longrightarrow$ 6CO$_2$ + 3H$_2$S + 6H$_2$O	-741	?

glucose whereas fermentative anaerobic organisms form only two. The oxidations of glucose by nitrate and sulphate as catalysed by dissimilatory anaerobic bacteria proceed with relatively more negative free energy changes than those catalysed by the fermentative anaerobes and here again these differences are reflected in the amount of ATP synthesized (Table 6). Individual dissimilatory organisms are fairly specific in the compounds they use as electron acceptors for the oxidation of glucose; the compound used and the extent to which it is reduced (e.g. $NO_3^- \longrightarrow NO_2^-$, N$_2$ or NH$_3$) varies with the enzyme complement of the organism and the environmental conditions. Although dissimilatory organisms are anaerobic, the reaction mechanisms of these organisms have more in common with aerobic organisms than the fermentative anaerobic organisms which do not require an external electron acceptor from the environment.

4.3   Oxidation of organic carbon and synthesis of ATP by heterotrophs

4.3.1   Anaerobic oxidations which do not involve an exogenous electron acceptor

The organisms which concern us in this section are those that catalyse the partial oxidation of glucose to pyruvate by the process of glycolysis but which are unable to oxidize pyruvate to CO$_2$. This process is referred to as fermentation. It occurs in many bacteria, yeasts and fungi when grown anaerobically. Certain tissues of aerobic organisms also exhibit fermentation when subjected to anaerobic conditions (e.g. muscle tissue during violent physical exercise and roots of plants when the soil becomes waterlogged). The essential features of glycolysis are summarized in Fig. 4–1. The oxidation reaction of glycolysis occurs in Phase 3 with NAD serving as the oxidizing agent. If cells catalysed the oxidation of phosphoglyceraldehyde by NAD without the incorporation of inorganic phosphate (P_i), the $\Delta G'$ would be extremely negative and much of the free energy of oxidation would be dissipated as heat. However, organisms possess an enzyme which catalyses the incorporation of P_i during the oxidation of phosphoglyceraldehyde to make a high energy phosphate bond in the product (diphosphoglycerate). This effectively conserves much of the free energy of oxidation in the product with the result that the $\Delta G'$ of the overall reaction is close to zero.

The free energy of oxidation of phosphoglyceraldehyde, conserved in diphosphoglycerate, is used for the synthesis of ATP in Phase 4 in a process known as a substrate level phosphorylation. As noted in section 1.5, the hydrolysis of diphosphoglycerate has a highly negative $\Delta G'$

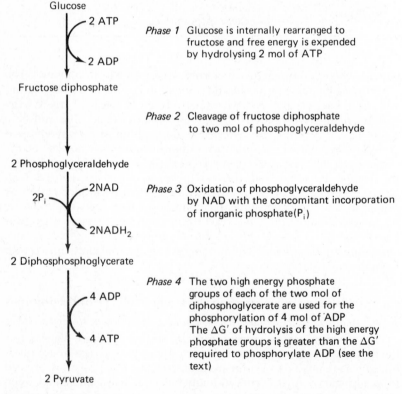

Fig. 4–1   Summary of the main component processes of glycolysis. Glucose is only one of several hexose (C_6) sugars which can be metabolized via the glycolytic pathway.

(reaction *(4.1)* which is more than sufficient to fulfil the positive free energy requirement for the phosphorylation of ADP (reaction *(4.2)*)):

(4.1)   Diphosphoglycerate $\longrightarrow$ Phosphoglycerate $+P_i$
$$(\Delta G' = -49.4 \text{ kJ mol}^{-1})$$
(4.2)   ADP $+P_i \longrightarrow$ ATP $(\Delta G' = +30.5 \text{ kJ mol}^{-1})$

(4.3)   Diphosphoglycerate $+$ ADP $\longrightarrow$ Phosphoglycerate $+$ ATP
$$(\Delta G' = -18.9 \text{ kJ mol}^{-1})$$

Organisms contain an enzyme which specifically couples reactions *(4.1)* and *(4.2)* together. It follows that the reaction catalysed by the enzyme (equation *(4.3)*) proceeds with a strongly negative free energy change. This results in a high yield of ATP at equilibrium. Organisms structurally rearrange the phosphoglycerate produced in equation *(4.3)* to form phosphopyruvate. This compound, like diphosphoglycerate, hydrolyses with a highly negative $\Delta G'$ (reaction *(4.4)*). Here again we find that organisms contain an enzyme which couples this reaction to the phosphorylation of ADP in a reaction with a highly negative $\Delta G'$ (equation *(4.6)*) and results in a very high yield of ATP at equilibrium.

(4.4)   Phosphopyruvate $\longrightarrow$ Pyruvate $+P_i$ $(\Delta G' = -61.9 \text{ kJ mol}^{-1})$
(4.5)   ADP $+P_i \longrightarrow$ ATP $(\Delta G' = +30.5 \text{ kJ mol}^{-1})$

(4.6)   Phosphopyruvate $+$ ADP $\longrightarrow$ Pyruvate $+$ ATP
$$(\Delta G' = -31.4 \text{ kJ mol}^{-1})$$

It follows that the oxidation of 2 molecules of phosphoglyceraldehyde supports the synthesis of 4 molecules of ATP. Since 2 molecules of ATP are consumed per molecule of hexose in Phase 1 of glycolysis (Fig. 4–1), and 2 molecules of phosphoglyceraldehyde are produced in Phase 2 it follows that glycolysis results in the net synthesis of 2 molecules of ATP per molecule of hexose oxidized.

The continued function of the glycolytic pathway shown in Fig. 4–1 requires a constant supply of ADP, P_i and NAD. The ATP produced from ADP and P_i is hydrolysed in a variety of energy-requiring processes in cells and the products (ADP and P_i) are returned for recycling in glycolysis. Similarly, mechanisms must exist for the oxidation of $NADH_2$ produced in glycolysis so that NAD can also recycle. Some of the $NADH_2$ is used for various reductive biosynthetic processes but this only accounts for a small proportion of the amount produced. Since the potential free energy of oxidation of $NADH_2$ by oxygen (-218 kJ mol^{-1}) cannot be realized for metabolic purposes under anaerobic conditions it follows that anaerobic organisms must use some other oxidizing agent which is available in substrate quantities for the oxidation of $NADH_2$. In fermentative anaerobes the pyruvate (carbon oxidation product) produced in glycolysis fulfils this role. Anaerobic organisms catalyse the oxidation of $NADH_2$ by pyruvate by one of several mechanisms, the most important being

(4.7) Pyruvate + $NADH_2$ ⟶ Lactate + NAD
(4.8) Pyruvate + $NADH_2$ ⟶ Ethanol + CO_2 + NAD

Lactate or ethanol plus CO_2 are therefore characteristic end products of glycolysis in fermentative anaerobic organisms and accumulate either in the organism or in the growing medium. Although hexoses (e.g. glucose) are initially oxidized to pyruvate the net free energy change for glucose oxidation in fermentative anaerobic organisms is measured between glucose and the eventual product, the latter being more reduced than pyruvate. The efficiency with which an anaerobic organism extracts energy from the oxidation of glucose to, say lactate, can be calculated by comparing the free energy change associated with the phosphorylation of 2 molecules of ADP (i.e. $2 \times +30.5 = +71$ kJ) with the value for the oxidation of glucose to 2 molecules of lactate $(-197$ kJ).

4.3.2   Aerobic oxidations

Aerobic organisms in the presence of oxygen catalyse the oxidation of organic carbon to CO_2. For hexoses, this entails the partial oxidation of hexose to pyruvate as discussed above for anaerobic organisms with the net production of 2 molecules of ATP and 2 molecules of $NADH_2$ per molecule of glucose oxidized. In aerobes pyruvate does not serve as the oxidant for the $NADH_2$ produced in glycolysis but is completely oxidized to CO_2 with the production of still more $NADH_2$ and other reducing equivalents. These in turn are oxidized by oxygen with the production of ATP as follows:

(4.9)

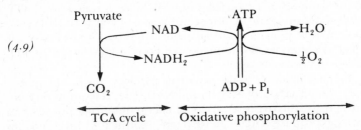

The oxidation of pyruvate to CO_2 with the production of $NADH_2$ (and other reducing equivalents) involves a series of reactions collectively known as the tricarboxylic acid cycle (TCA cycle). The oxidation of $NADH_2$ by oxygen with the associated phosphorylation of ADP is known as oxidative phosphorlyation and involves the electron transport chain (ETC). In aerobic eukaryotic organisms, these processes occur within the mitochondria of cells but in prokaryotic organisms no specialized organelle for this purpose exists.

The oxidation of pyruvate to 3 molecules of CO_2 by the TCA cycle is summarized in Fig. 4-2. Details of the processes involved can be found in

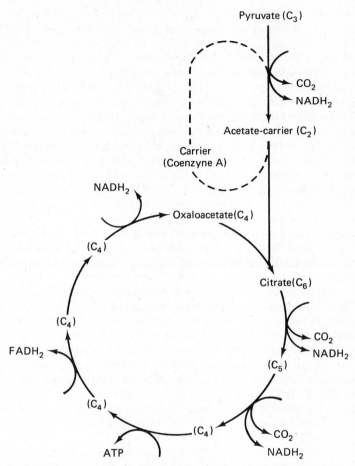

Fig. 4-2   Summary of the principal processes of the TCA cycle.

other books in this series. In summary, one molecule of pyruvate is oxidized by NAD (thereby forming $NADH_2$) with the elimination of one molecule of CO_2. The acetate so formed is attached to a carrier molecule (coenzyme-A). The acetyl moiety (C_2) is condensed with a C_4 molecule (oxaloacetate) to form a C_6 molecule (citrate) which is then progressively oxidized in a series of enzyme-catalysed reactions to form oxaloacetate (C_4) with the loss of two molecules of CO_2 $(2 \times C_1)$. The oxaloacetate produced replenishes the oxaloacetate consumed earlier in the cycle. A further three molecules of NAD and one molecule of FAD are reduced during the oxidation of citrate to oxaloacetate. Also associated with this process is the phosphorylation of one molecule of ADP by a complex mechanism involving substrate level phosphorylation. Hence the net products per molecule of pyruvate (C_3) fed into the TCA cycle are $4NADH_2$, $1FADH_2$, 1 ATP and $3CO_2$ ($3 \times C_1$) with no net change in the level of oxaloacetate. These values are doubled (Fig. 4–3) per molecule of hexose (which yields two molecules of pyruvate).

In fermentative anaerobes the $NADH_2$ produced in glycolysis is oxidized by pyruvate. In aerobic organisms, however, pyruvate is fed into the TCA cycle and some other mechanism must exist for the oxidation of $NADH_2$ produced in the cytoplasm. In eukaryotes the reducing equivalents in $NADH_2$ in the cytoplasm are transferred to mitochondria. However, like the chloroplast envelope, the inner mitochondrial membrane is impermeable to $NADH_2$. Various metabolites (e.g. α-glycerophosphate), freely permeable to the mitochondrial membrane, serve as carriers to transport the reducing equivalents into the mitochondrion (Fig. 4–3) in a manner analogous to the shuttle mechanism described for the outward transport of reducing equivalents from chloroplasts (Fig. 3–3). The reducing equivalents transported from cytoplasmic $NADH_2$ into mitochondria via the shuttle mechanism reduce FAD.

In eukaryotes, the main effect of the oxidation reactions of the TCA cycle and glycolysis (in conjunction with shuttle mechanisms) is the conservation of a large proportion of the potential free energy of oxidation of the carbon source as reduced forms of biological reducing agents ($NADH_2$ and $FADH_2$) in mitochondria. As noted previously these compounds can be oxidized in mitochondria by molecular oxygen with a highly negative free energy change, the value for $NADH_2$ being somewhat greater than $FADH_2$ (Table 3). However, electron flow from $NADH_2$ and $FADH_2$ to oxygen (i.e. oxidation of $NADH_2$ and $FADH_2$) does not occur in a single step. Embedded in the inner membrane of mitochondria are a series of complex compounds, referred to as redox carriers, each with a characteristic redox potential. The most important of these are a set of closely related proteins called cytochromes with substituent iron-porphyrin groups. Collectively they form the electron transport chain (ETC). The cytochromes and the other redox carriers are arranged in the inner membrane in such a way that electron flow from $NADH_2$ to oxygen proceeds successively from $NADH_2$ to the redox carrier with the next most negative redox potential to the carrier with the next most negative redox potential and so on until the electron is transferred to oxygen to form water (Fig. 4–4). In this way, electron flow proceeds in a series of reactions, each with a relatively small free energy change. Electron flow from $NADH_2$ to oxygen involves three reactions with $\Delta G'$ values in excess of -38 kJ but for $FADH_2$ which enters the mainstream of electron flow at cytochrome-b (Fig. 4–4), there are only two such reactions. Electron flow from $NADH_2$ and $FADH_2$ ($2e^-$) to oxygen is associated with the phosphorylation of three and two molecules of ADP respectively. This process which is known as oxidative phosphorylation, is quite complex

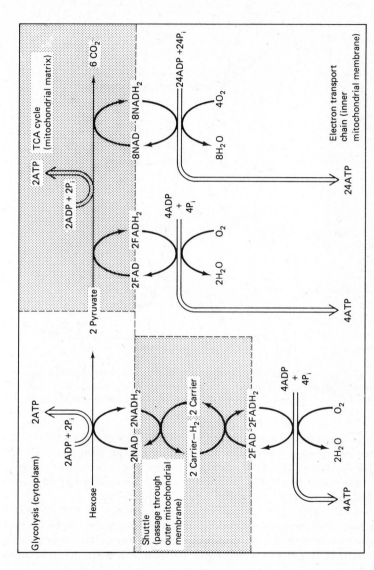

Fig. 4–3 Summary of the major processes involved in the oxidation of a molecule of hexose in a eukaryotic aerobic organism.

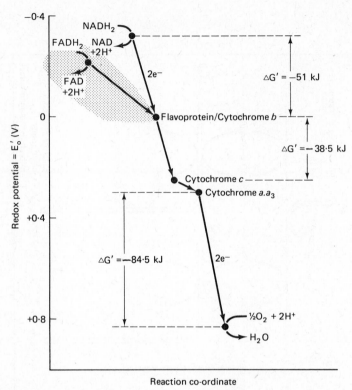

Fig. 4–4 Redox potentials of the main component redox carriers of the electron transport chain of mitochondria. The free energy changes associated with the passage of the two electrons between some of the components are also shown.

and here again readers are referred to the article by HINCKLE and McCARTY (1978). Nevertheless it is evident that the free energy changes associated with the oxidation of $NADH_2$ and $FADH_2$ by oxygen (−218 and −199 kJ mol^{-1}) are more than sufficient to account for the synthesis of three and two molecules of ATP (+91.5 and +71 kJ) respectively.

From the foregoing information, the total amount of ATP produced per molecule of hexose oxidized by an aerobic organism can be derived as shown overleaf. It is apparent that an aerobic organism can extract 18 times more ATP per molecule of glucose than a fermentative anaerobic organism. Compared with the free energy oxidation of glucose by oxygen (−2870 kJ mol^{-1}), approximately $36 \times -30.5 = -1098$ kJ are conserved as ATP.

The mechanisms for the oxidation of hexoses to CO_2 and the synthesis of ATP by aerobic prokaryotic organisms are analogous to those described above for the mitochondria of eukaryotes. In prokaryotes,

Process	Molecules of ATP synthesized
Glycolysis (1 hexose ⟶ 2 Pyruvate)	
Substrate level phosphorylation	2
2 $NADH_2$ transported via shuttle to mitochondria	
⟶ 2 $FADH_2$ (oxidized via ETC) ⟶ 2 × 2 ATP	4
TCA cycle (2 Pyruvate ⟶ $6CO_2$)	
Substrate level phosphorylation	2
8 $NADH_2$ (oxidized via ETC) ⟶ 8 × 3ATP	24
2 $FADH_2$ (oxidized via ETC) ⟶ 2 × 2ATP	4
	36 ATP

however, the entire cell can be likened to a mitochondrion. The reactions of glycolysis and the TCA cycle occur in the matrix of the cell. The components of the electron transport chain of prokaryotes differ in details from those of eukaryotes and are located in the plasma membrane.

4.3.3   Oxidations coupled to oxidized forms of nitrogen

Several groups of bacteria catalyse the complete oxidation of organic carbon to CO_2 under anaerobic conditions. Typical of this group are *Pseudomonas* and *Paracoccus denitrificans* which have a glycolytic pathway and a TCA cycle for the oxidation of hexoses as described for anaerobic prokaryotes but, under anaerobic conditions, use oxidized forms of nitrogen (e.g. nitrate) as the terminal electron acceptor for the oxidation of $NADH_2$ and $FADH_2$ (i.e. dissimilatory nitrate reduction). It follows that these organisms consume nitrate (and consequently produce reduced forms of nitrogen) at rates approaching those for the consumption of organic carbon. One such process is the reduction of nitrate to gaseous nitrogen, a process known as denitrification. Organisms catalysing dissimilatory nitrate reduction are characteristically found in anaerobic environments (e.g. waterlogged soils, muds and marshes) containing relatively high concentrations of oxidized forms of nitrogen and organic carbon.

Under aerobic conditions electron flow from $NADH_2$ to oxygen in *Paracoccus denitrificans* proceeds via an electron transport chain similar to that described in mitochondria. However, under anaerobic conditions in the presence of nitrate or nitrite the bacterium synthesizes additional redox carriers. In the presence of these adaptive carriers electrons flow from $NADH_2$ to nitrate and/or nitrite. The passage of electrons from $NADH_2$ to nitrite (thereby forming N_2 gas) is associated with the phosphorylation of only 2 molecules of ADP per molecule of $NADH_2$ oxidized.

4.3.4   Oxidations coupled to oxidized forms of sulphur

Bacteria belonging to the genera *Desulfovibrio* and *Desulfotomaculum* are anaerobic organisms which oxidize organic carbon and use oxidized forms of sulphur as terminal electron acceptors (i.e. dissimilatory sulphate reduction). They are usually restricted to anaerobic habitats rich in organic carbon and oxidized forms of sulphur (e.g. marine muds, soil, oil-bearing environments). The mechanisms of oxidation of carbon substrates (usually containing three or four carbon atoms) in *Desulfovibrio* and *Desulfotomaculum* are not well understood. However, it appears that oxidized forms of ferredoxin or flavodoxin serve as intermediate electron acceptors and that electron flow from reduced forms of these compounds to sulphate proceeds via cytochrome-c_3 with the concomitant synthesis of ATP. The reduced forms of sulphur produced by dissimilatory sulphate reducing bacteria (e.g. H_2S) largely account for the vile stench associated with anaerobic environments.

In summary, the process of respiration involves the partial or complete oxidation of an organic carbon source in a reaction with an overall negative free energy charge. The free energy from these reactions is trapped as ATP which in turn is used as the energy source for most other cellular activities. A diversity of compounds (e.g. oxygen, nitrate, sulphate) are used as electron acceptors by different organisms. In the absence of a suitable oxidizing agent from the environment a carbon oxidation product (e.g. pyruvate) can be used as electron acceptor (fermentation). All cellular organisms, including autotrophs, exhibit respiration but respiration is the only mechanism for the synthesis of ATP in heterotrophs.

5 Interdependence of Organisms for Carbon, Nitrogen, Sulphur and Energy

5.1 General principles

It is apparent from the preceding chapters that individual organisms require carbon, nitrogen, sulphur and energy in different forms. Plants for example use light energy for the synthesis of biological matter from CO_2 and inorganic salts. Conversely heterotrophs derive their energy from the oxidation of biological matter synthesized by autotrophs though heterotrophs also use a portion of their organic carbon source for the synthesis of new biological matter. In addition to the dependence of heterotrophs on autotrophs for a supply of organic carbon, autotrophs, as noted in Chapter 1, are equally dependent on the oxidation of organic carbon to CO_2 by heterotrophs, i.e. the two groups are interdependent with respect to carbon. Similarly, organisms are interdependent in their requirements for nitrogen and sulphur. The interdependence of organisms associated with aerobic terrestrial environments with respect to their supplies of carbon, nitrogen and sulphur and their dependence on an environmental source of energy are summarized in Fig. 5–1. This cycle can be maintained so long as there is an input of light energy. Energy, unlike the individual elements, does not recycle within the ecosystem. Light energy from the environment is stored as chemical energy in biological matter by plants. Collectively, a series of heterotrophs (e.g. Fig. 1–1) oxidize biological matter to CO_2 and dissipate the chemical energy back into the environment as heat energy. However, in some anaerobic ecosystems cycling of nitrogen and sulphur bear little resemblance to the scheme shown for aerobic ecosystems in Fig. 5–1. This is because nitrogen and sulphur, in addition to being required for the synthesis of some essential biological compounds, also participate as electron acceptors and donors in the electron transport chains of particular organisms (see Chapter 4). In these organisms, transformations of nitrogen and sulphur are stoichiometrically equivalent to carbon and therefore assume similar importance. Some examples are discussed later in this chapter.

Cycling of carbon, nitrogen and sulphur and the flow of energy between organisms is a function of the metabolic processes of the individual organisms within a community. Global cycling of the elements is obtained by summing the various transformations catalysed by organisms with the gains and losses from the biosphere by non-biotic processes. The latter factors include additions to the biosphere by rock weathering, geological events and the activities of man (e.g. application of

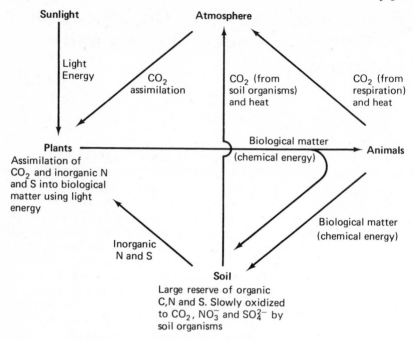

Fig. 5–1 Energy flow and biological cycling of carbon, nitrogen and sulphur in a terrestrial ecosystem involving green plants as the primary producers.

fertilizers, burning fossil fuels, etc.) and removal from the biosphere by the action of wind, water and deposition in sediments (e.g. coal formation, marine sulphide deposits, etc.). Details of global cycling of individual elements can be found in other books in this series. The remainder of this chapter, however, is concerned with examining the interdependence of organisms within specific ecosystems with respect to their requirements for carbon, nitrogen, sulphur and energy and any net inputs and outputs that might be involved. They demonstrate the diversity of interrelationships exhibited by organisms in ecosystems, including some ecosystems in which autotrophs other than green plants act as the primary producers.

5.2 Aerobic ecosystems with green plants as the primary producers

Most of the terrestrial surface of the earth and the oceans are aerobic. Green plants occur throughout the aerobic regions of the earth except in those areas where some other factor such as lack of water (e.g. deserts) or light (e.g. below 200 metres in the oceans) prevents plant growth. Plants, in association with the heterotrophic organisms that they support, account for most of the biological cycling of carbon, nitrogen and sulphur. In terrestrial ecosystems, the conserved chemical energy in the organic compounds synthesized by plants is used for the growth and function of animals and soil microorganisms and results eventually in the oxidation of organic carbon, nitrogen and sulphur to oxidized inorganic forms of these elements though at any point in time a substantial proportion of all three elements is present in soil organic matter. The oxidized forms are recycled by plants. The time taken for all of the biologically available forms of an element to undergo one complete turnover is referred to as the cycling time. It varies between elements and with the soil fertility. For the terrestrial system, the cycling time for carbon is approximately 300 years and is probably fairly uniform over most of the earth's surface since carbon is recycled via the atmosphere and mixed by air turbulence. For sulphur deficient soils cycling times can be as short as seven years but in sulphur sufficient soils with a larger total amount of sulphur, cycling times are appreciably longer. It follows that the cycle would cease in the absence of sunlight.

Cycling of carbon within an aerobic ecosystem involves various non-biological components in addition to organisms. The non-biological components include CO_2 and some materials of biological origin such as organic matter and detritus. The amount of carbon associated with each of the biological and non-biological components can be determined by fractionating them and analysing them for carbon. Within an ecosystem, carbon flows from CO_2 to autotrophs to various biological and non-biological components (e.g. heterotrophs and organic matter). The fluxes of carbon between the components of an aerobic ecosystem can be determined by supplying $^{14}CO_2$ to the ecosystem and quantitatively monitoring the rate of ^{14}C-labelling of the autotrophs (primary producers) and, over longer periods of time, subsequent labelling of other organisms and non-biological components. In practice, carbon flux by the heterotrophs in an ecosystem is more usually studied by supplying a ^{14}C-labelled organic compound (e.g. [^{14}C] glucose). Studies of carbon flux are most readily determined in aquatic rather than terrestrial ecosystems since aquatic systems are easier to sample and the carbon fluxes can be determined using labelled compounds which are water-soluble and therefore readily available to organisms. The results of a study of the distribution and flux of carbon in the top metre of water in an aerobic lake for a single 24-hour period in summer are shown in Fig. 5–2. The data show that most of the carbon in the lake is present in inorganic forms (dissolved CO_2, HCO_3^-, etc.) and that carbon flows from this component through algae (primary producers) to the heterotrophs (zooplankton and bacteria) via a complex web of fluxes. It is also evident that dissolved organic carbon (from algae) and particulate organic matter of biological origin (detritus) form important reservoirs of carbon for heterotrophic organisms. The data also show that the amount of carbon assimilated by the algae in a single day is about 60% of the carbon present

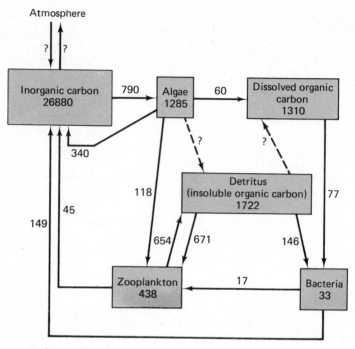

Atmosphere

Fig. 5–2 Cycling of carbon between the biological and non-biological components in the top metre of an aerobic aquatic ecosystem associated with a freshwater lake (Frains Lake, Michigan, U.S.A.). Values within the blocks represent the amount of carbon associated with the component in μg of carbon per litre of water. Values beside the arrows denote the carbon flux between the components in μg l^{-1} day^{-1}. The values apply for the 24-hour period of a specific day in the summer of 1968. (Re-drawn from Saunders, G. W. in CAIRNS, 1977.)

in the algae. This implies that the average time of retention of carbon in algal cells before being subject to respiration by the algae or entering the latter part of the food chain as organic carbon is only 1.7 days. Further, by comparing the amount of carbon assimilated per day with the amount of inorganic carbon in the lake it can be calculated that the algae would exhaust the supply of inorganic carbon in about 30 days unless the supply was constantly renewed by the respiratory activities of organisms in the lake and dissolving of CO_2 from the atmosphere. Some organic carbon fluxes between the heterotrophs in the lake. This is because the two groups of heterotrophs (bacteria and zooplankton) are comprised of populations of different species which act as primary, secondary, tertiary, etc., consumers of organic carbon. Thus organic carbon could flow from a primary consuming bacterium, to a secondary consuming zooplankton, to a tertiary consuming bacterium. Clearly, the amount of carbon transferred at each stage would decrease as each organism oxidizes a portion of the organic carbon.

The sum of the daily carbon fluxes from organisms to inorganic carbon amounts to 68% of the CO_2 assimilated by the algae. The carbon cycle in the top metre of the water on the particular day the lake was analysed was therefore not balanced. The remaining 32% could accumulate as dissolved organic carbon and detritus thereby causing an increase in the heterotroph population at some later date thereby causing the flux of organic carbon to CO_2 to exceed CO_2 consumption. Alternatively, a certain amount of detritus could settle to the floor of the lake. Depending on the prevailing conditions the precipitated carbon could be oxidized by heterotrophic organisms associated with the bottom of the lake or simply accumulate, thereby forming a potential source of fossil carbon (and energy) for the future. The latter process effectively removes carbon from biological circulation. The carbon fluxes between the various biological and non-biological components of an ecosystem and the amount of carbon associated with them will vary throughout the year due to changes in environmental and nutritional conditions. Consquently the relative numbers and metabolic activities of individual species in the ecosystem will also change and lead to the introduction of new biological and non-biological components in the carbon cycle.

Within an aerobic ecosystem the autotrophic and heterotrophic organisms which reduce and oxidize carbon respectively can play quite unrelated roles in the reduction and oxidation of nitrogen and sulphur. Many heterotrophic organisms for example use oxidized forms of nitrogen and sulphur for the synthesis of organic molecules which contain these elements in reduced forms. Thus, whereas heterotrophs are dependent on autotrophs for their supply of carbon (and therefore energy), certain heterotrophs compete with autotrophs for oxidized forms of nitrogen and sulphur. It therefore follows that the biological and non-biological components of the nitrogen and sulphur cycles in a given aerobic ecosystem may bear little relation to those of the carbon cycle.

The input of energy into an aerobic ecosystem involving plants as the primary producers begins with the absorption of incoming solar radiation by the light-absorbing pigments of plants. In a typical terrestrial ecosystem approximately 0.2–0.3% of the incoming solar energy accumulates as chemical energy (organic carbon, nitrogen, etc.) in plant matter. The flow of energy then follows essentially the same pathway as organic carbon through a sequence of heterotrophs and non-biological components (e.g. Fig. 5–2). The flow of energy ceases with the complete oxidation of organic carbon to CO_2 though the deposition of organic carbon in sediments, which can form a source of energy for men's domestic and industrial requirements in the future, is an important exception.

The total amount of energy (E_t) consumed by a heterotroph in an ecosystem can be determined from the expression $E_t = E_r + E_y$ where E_r denotes the energy lost by respiration and E_y the net gain or loss of stored energy (i.e. biological matter) by the organism. E_r can be determined by measuring respiration and applying expressions such as those shown in equations *(1.1)* and *(1.2)* relating CO_2 evolution to free energy changes. E_y can be ascertained by determining the changes in the mass of dry matter of the species over a given period of time. The chemical energy associated with a given amount of dry matter can be determined by oxidizing it in a bomb calorimeter thereby determining the energy equivalent (joules per gram of dry matter). The product of the change in mass of dry matter per unit time and energy equivalent gives a measure of E_y. This can be determined for each heterotroph or groups of heterotrophs within an ecosystem. A similar approach is used to determine the net amount of energy fed into an ecosystem by the primary producers and the amount of energy channelled through detritus or introduced from neighbouring ecosystems. Since some individuals are subject to mortality and some species exhibit relatively short life cycles, possibly shorter than the period between taking samples for analysis, corrections must be made for the amount of biological matter that they consume or produce. Other corrections are also necessary.

The results of a study of energy flow in the heterotrophic organisms within an aerobic ecosystem are shown in Fig. 5–3. The values shown apply to a spring covering an area of 116 m². Water oozes up uniformly over this area covering it to a depth of about 3 cm. In this ecosystem, three plants (*Lemma, Impatiens* and *Bacopa*) account for the bulk of the primary production. These, together with net inputs of biological matter such as leaves, twigs, fruits, etc., from surrounding ecosystems contribute to a reservoir of detritus which serves as a source of organic carbon for the primary heterotrophs. Only a relatively small proportion of the energy flowing into the primary heterotrophs is conserved; the remainder is lost to the environment during oxidation of organic carbon to CO_2. For example, only 2766 of the 9975 kJ m^{-2} year^{-1} (i.e. 28%) of the energy consumed by the primary heterotrophs is available for consumption by subsequent organisms. This pattern is repeated by the secondary and tertiary organisms respectively.

5·3   The Galapagos Rift ecosystem

In the Pacific Ocean between the Galapagos Islands and the mainland of South America is a a rift valley some 2500 m below sea level known as the Galapagos Rift. The ambient temperature of the water in the bottom of the rift is about 2°C. Located within the rift is a small ridge rising about 60 m above the floor of the valley. Along the ridge are a series of vent areas about 30–100 m wide containing a series of small fissures about 2–10 cm across. Water escapes from these fissures (vented water) at temperatures of about 5–15°C greater than the ambient water. The water

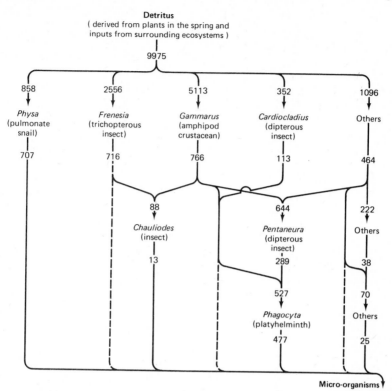

Fig. 5–3 Energy flow between the heterotrophic organisms associated with a shallow spring in Iowa, U.S.A. All values are express in kJ m⁻² year⁻¹. (Data abstracted from Tilly, L. S. (1965). *Ecological Monographs*, **38**, 169–97. Copyright 1965 by the Ecological Society of America.)

temperature above the vent areas decreases with increasing height above the vent area; at 180 m, the temperature is essentially the same as the ambient water (2°C). From an analysis of the silica content and temperature of the vented water it has been suggested that ambient water percolating downwards is heated by hot rocks at a temperature greater than 300°C. The much lower temperature of water emanating from the fissures is due to mixing of the hot water as it rises with cold ambient water below the surface of the vent areas (Fig. 5–4).

Sunlight does not penetrate through the 2500 m of water overlying the vent areas. Nevertheless, the vent areas support rich and unique communities of organisms characterized by various clams, mussels and limpets, most of which are distinct from their shallow-water counterparts. The areas adjacent to the vents contain no such prolific growth of organisms. It is evident that the organic metabolites required

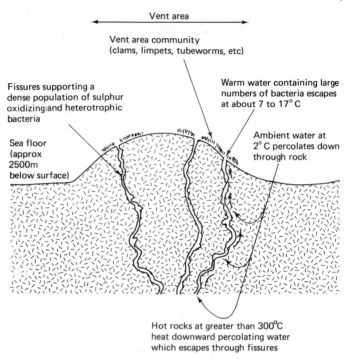

Vent area

Vent area community
(clams, limpets, tubeworms, etc)

Fissures supporting a
dense population of sulphur
oxidizing and heterotrophic
bacteria

Warm water containing large
numbers of bacteria escapes
at about 7 to 17°C

Sea floor
(approx
2500m
below surface)

Ambient water at
2°C percolates down
through rock

Hot rocks at greater than 300°C
heat downward percolating water
which escapes through fissures

Fig. 5–4 Some features of the submarine thermal springs and vent areas of the Galapagos Rift. The vent areas are located approximately 2500 m below the ocean surface and receive no significant sunlight. (Compiled from the data of CORLISS *et al.*, 1979.)

by the animals of the vent areas cannot originate from photosynthetic organisms and must be synthesized by some other means. The warm water issuing from the fissures (vented water) contains H_2S at concentrations of 20–160 μM and sulphur-oxidizing autotrophic bacteria and heterotrophic bacteria estimated at 0.1–1 g of organisms per litre. It is likely that the temperature and the concentration of H_2S within the fissures is even greater than those of the vented water. These conditions are favourable for the growth of sulphur-oxidizing autotrophic bacteria. Presumably crevices within the fissures afford a niche for their growth and a proportion of them are sloughed out of the crevices by upward movement of vent water and carried to the vent areas. It is likely that the entire community of organisms of the vent areas, including the larger animals, is dependent on the production of biological matter by chemoautotrophs which derive their energy for CO_2 assimilation from the oxidation of H_2S produced by geological activity below the fissure. Perhaps it is significant that most of the animals found in the communities of the vent areas are filter feeders (i.e possess mechanisms for filtering out small particulate matter such as bacteria from relatively large volumes of water). At present little is known about the recycling of carbon, nitrogen and sulphur in vent area ecosystems.

It follows that the continued existence of the vent area communities is dependent on the upward currents of warm water and a supply of H_2S as a source of energy for the synthesis of biological matter by chemoautotrophs. In this respect it is of interest that areas are found on the ridge of the Rift Valley which do not currently exhibit venting of warm water or support a vent community but which appear to have supported a vent community previously. These areas are characterized by specific geomorphological features and piles of decaying clam shells characteristic of vent areas and the communities that they support. Radiometric dating of the decaying shells indicates that they are 10–20 years old suggesting that the vent areas might be quite short lived and rapidly changing.

5·4   Aquatic ecosystems involving an anaerobic component

Aerobic organisms require nitrogen and sulphur for incorporation into biological matter. Relative to carbon, nitrogen and sulphur are required in relatively small amounts (Table 1) and this is reflected in the fluxes of these elements in aerobic ecosystems. However, many mixed aerobic/anaerobic ecosystems contain high populations of organisms whose nitrogen and sulphur metabolism is stoichiometrically equivalent to carbon metabolism (e.g. organisms catalysing dissimilatory nitrate and sulphate reduction). It follows that the nitrogen and sulphate fluxes relative to carbon are much greater in mixed aerobic/anaerobic ecosystems and that the fluxes of all three elements are intrinsically related.

Mixed aerobic/anaerobic ecosystems can be found in many ponds and marshes and even in large bodies of water such as the Black Sea. The formation of these ecosystems is usually initiated by the introduction of large amounts of organic matter and/or the inorganic salts required for the growth of autotrophs. If the amounts of these materials relative to the volume of water are small then the conditions are favourable for the balanced physiological activity of aerobic autotrophs (utilizing the salts, light and CO_2) and aerobic heterotrophs which decompose the incoming organic matter and the biological matter produced by the photo-autotrophs. However, if the body of water suddenly receives a large influx of biological matter (e.g. leaf litter from deciduous trees in the autumn) this can cause the physiological activities of the photoautotrophs and aerobic heterotrophs to become out of balance. The influx of organic matter creates conditions favourable for the growth of heterotrophs which consume oxygen faster than it is produced by aerobic photoautotrophs and dissolved from air at the surface of the pond. This results in the depletion of oxygen in the water and induces anaerobic conditions. Alternatively, if the water body receives a large influx of inorganic salts (e.g. run-off water from surrounding countryside which has been treated with fertilizers containing nitrogen, sulphur and phosphorus), this induces vigorous growth of photoautotrophs in the water. The biological matter that they produce leads to a massive increase in the population of heterotrophs with the consequent onset of anaerobic conditions. Aerobic photoautotrophs and heterotrophs cease growing under these conditions.

In Chapter 4 it was noted that the energy-generating mechanisms of anaerobic organisms were essentially of two types, i.e. the dissimilatory organisms which use inorganic compounds other than oxygen as terminal electron acceptor (e.g. nitrate, sulphate) and the fermentative organisms which use a partially oxidized carbon metabolite (e.g. pyruvate) produced by the organism itself. It was also noted from a consideration of the free energy changes associated with these reactions (Table 6) and the ATP-generating mechanisms associated with them that dissimilatory organisms synthesized more ATP from a given organic substrate than fermentative organisms. This therefore places the dissimilatory anaerobic organisms at a selective advantage if the supply of substrate is limiting. It therefore follows that in the absence of oxygen, inorganic nitrate and sulphate are readily reduced by dissimilatory organisms. The relative rates of anaerobic oxidation of organic matter (and hence growth) by fermentative and dissimilatory organisms depends on the availability of inorganic nitrogen and sulphur. Anaerobic bodies of water containing large amounts of these elements commonly contain high concentrations of the dissimilatory organisms which require them in large quantities as electron acceptors.

Once anaerobic conditions have been initiated a population of organisms develops in which individual organisms are interdependent for a supply of appropriate forms of carbon, sulphur and, to a lesser extent, nitrogen. For example, in a body of water containing sulphate and sedimented dead biological matter, dissimilatory sulphate-reducing organisms oxidize the biological matter, producing reduced forms of sulphur (e.g. H_2S). Higher in the profile (where sunlight can penetrate) H_2S can be used as an electron donor for the light-dependent electron-transport mechanisms of anaerobic photosynthetic bacteria (Fig. 5–5) thereby incorporating CO_2 into organic matter with the concomitant oxidation of H_2S to elemental sulphur (see section 2.1). The organic carbon they produce can be recycled via the dissimilatory and fermentative heterotrophs. The species of organisms present near the surface of the water depends on the partial pressure of oxygen. In all bodies of water, oxygen is dissolved from the atmosphere creating relatively aerobic conditions. However, especially in small ponds which receive little mixing by wave turbulence, the layer of aerobic water may

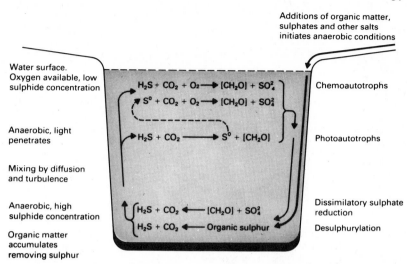

Fig. 5–5 Biological cycling of carbon and sulphur in an anaerobic body of water containing relatively high concentrations of sulphur. The dissolution of oxygen from air at the surface of the water creates a thin aerobic layer of water at the surface.

be so thin and the oxygen concentration so low that aerobic photoautotrophs are unable to grow. However, chemoautotrophs grow readily at low partial pressures of oxygen. The upward diffusion of reduced forms of nitrogen and sulphur (produced by organisms lower in the water) into the aerobic layer provides conditions suitable for the growth of chemoautotrophs. These organisms (e.g. *Thiobacillus, Nitrosomonas*) catalyse the aerobic oxidation of reduced forms of nitrogen, sulphur and other elements in exergonic reactions which are used as energy sources for the assimilation of CO_2 into biological matter. In addition to cycling organic carbon back to heterotrophs, the chemoautotrophs by their very nature produce large amounts of oxidized forms of nitrogen and sulphur. With the exception of gaseous nitrogen these can be recycled and used as terminal electron acceptors by dissimilatory organisms lower in the water profile.

The degradation of organic matter in a mixed aerobic/anaerobic environment can be studied in the laboratory in an artificial ecosystem. In one such experiment a tank was almost filled with a sediment of sand mixed with chopped up leaves of eelweed (*Zostera*), the latter acting as a source of organic carbon. This is referred to as the sediment. The tank was then filled with aerated sea water which in addition to serving as the liquid phase also provided an inoculum of organisms normally associated with sea water. Aerated sea water was continually cycled over the surface of the sediment (Fig. 5–6). The tank was illuminated with a light source and a

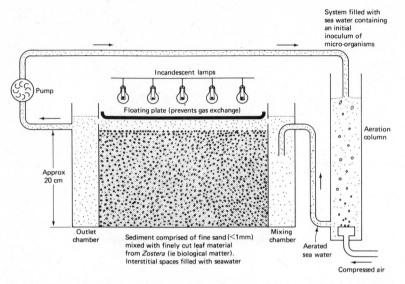

Fig. 5–6 Construction of an experimental system for the study of carbon and sulphur cycling by marine micro-organisms in a mixed aerobic/anaerobic eco-system (see the text). This model can also be used to study the succession of organisms as the sediment becomes anaerobic and organic matter in the sediment is oxidized. (Re-drawn from Jorgensen, B. B. and Fenchel, T. (1974). *Mar. Biol.*, **24**, 189–201.)

floating cover placed on the surface of the water to prevent the uncontrolled entry of oxygen from the atmosphere. Aerated sea water contains relatively high concentrations of sulphur as sulphate which percolates down into the sediment. The fate of the sulphate in the tank was followed by periodically removing samples of sediment or sea water (containing organisms and non-biological material) and incubating them with very small samples of sulphate labelled with the radioactive isotope ^{35}S. On concluding the incubation the various biological and non-biological components were separated and the amount of ^{35}S and the form in which it was present (e.g. S^{2-}, SO_4^{2-}, S^0, organic compounds) determined. In this way it was demonstrated that sulphate was reduced to H_2S by dissimilatory sulphate-reducing bacteria (e.g. *Desulfovibrio*) in an anaerobic zone in the sediment. Some of the H_2S produced precipitated non-biologically as FeS. The remainder was re-oxidized to elemental sulphur and sulphate by anaerobic photosynthetic bacteria and chemoautotrophs near the surface of the sediment. Knowing the amount of oxygen consumed by the sediment (47.0 μmol cm^{-3} of sediment) and the amount of sulphate reduced to sulphide (13.9 μmol cm^{-3}) together with the stoichiometries relating the oxidation of carbohydrate by oxygen and sulphate (see Table 6), it was calculated that the amount of organic carbon oxidized by aerobic and dissimilatory sulphate-reducing organisms amounted to 1.4 and 0.84 mg cm^{-3} respectively. Thus the dissimilatory organisms played a very significant part in the breakdown of organic matter in this experiment. Furthermore, the sulphide they produced was used by the sulphide-oxidizing photoautotrophs and chemoautotrophs to synthesize 1.1 mg organic matter cm^{-3}. It is therefore evident that carbon is also recycled by this system though it is not totally self sustaining.

The foregoing examples demonstrate that once a community of organisms is established in a mixed aerobic/anaerobic environment receiving an external source of energy and containing high concentrations of carbon, nitrogen and sulphur, then the community is largely self sustaining with respect to its requirements for oxidized and reduced forms of carbon, nitrogen and sulphur. In practice these communities also require constant additions of carbon (CO_2 or organic) to replace organic carbon deposited at the base of the pond. Small inputs of organic and/or inorganic nitrogen and sulphur are also required to replace nitrogen lost as N_2 (gas) by denitrifying bacteria and sulphur lost by precipitation as insoluble sulphides or volatilized as H_2S. Some nitrogen and sulphur is also deposited in organic matter. The external source of energy can take one or more forms. Probably the most important of these is light which can be utilized by anaerobic photoautotrophs in the anoxic zone and in some instances by aerobic photoautotrophs at the surface. Additions of organic matter from neighbouring ecosystems also serve as an energy source. In addition, chemoautotrophs synthesize biological matter from CO_2 thereby supplying energy. This process however consumes oxygen. Thus ecosystems with an anaerobic component, unlike aerobic ecosystems in which plants serve as the primary producers, exhibit net consumption of oxygen. In effect oxygen can be considered as an energy source though in chemical terms the energy is only made available when used by chemoautotrophs to oxidize reduced forms of sulphur and nitrogen.

Further Reading

Bioenergetics

BAKER, J. J. W. and ALLEN, G. E. (1970). *Matter, Energy and Life*, 2nd edition. Addison-Wesley, Reading Mass. and London.

KLOTZ, I. (1967). *Energy Changes in Biochemical Reactions*. Academic Press Inc., New York and London.

LEHNINGER, A. (1971). *Bioenergetics*, 2nd edition. W. A. Benjamin Inc., New York.

Energy generating mechanisms of organisms and assimilation of inorganic compounds

ALEEM, M. I. H. (1970). Oxidation of inorganic compounds of nitrogen. *Annu. Rev. Plant Physiol.*, **21**, 67–90.

BENEDICT, C. R. (1978). Nature of obligate photoautotrophy. *Annu. Rev. Plant Physiol.*, **29**, 67–93.

HINCKLE, P. C. and McCARTY, R. E. (1978). How cells make ATP. *Scient. Am.*, **238**, 104–23.

KIRK, J. T. O. and TILNEY–BASSETT, R. A. E. (1978). *The Plastids*, 2nd edition. Elsevier/North Holland/Biomedical, Amsterdam, New York and Oxford.

LE GALL, J. and POSTGATE, J. R. (1973). The physiology of sulphate reducing bacteria. *Adv. Microbial Physiol.*, **10**, 81–133.

SOKATCH, J. R. (1969). *Bacterial Physiology and Metabolism*. Academic Press, London and New York.

Cycling of elements and energy flow

CAIRNS, J. (Ed.) (1977). *Aquatic Microbial Communities*. Garland Publishing Inc., New York and London.

CORLISS, J. B. *et al.* (1979). Submarine thermal springs on the Galapagos Rift. *Science*, **203**, 1073–83.

NIELSEN, D. R. and MacDONALD, J. G. (Eds) (1978). *Nitrogen in the Environment*. Academic Press, New York.

NRIAGU, J. O. (Ed.) (1976). *Environmental Biochemistry*, Vol. 1. *Carbon, Nitrogen, Phosphorus, Sulphur and Selenium Cycles*. Ann Arbor Science, Ann Arbor, Michigan.

ODUM, E. P. (1971). *Fundamentals of Ecology*, 3rd edition. W. B. Saunders Co., Philadelphia.

The following booklets from the series *Studies in Biology* (Edward Arnold, London) are also relevant:

ANDERSON, J. W. (1978). *Sulphur in Biology*, no. 101.

HALL, D. O. and RAO, K. K. (1977). *Photosynthesis*, 2nd edition, no. 37.

PHILLIPSON, J. (1966). *Ecological Energetics*, no. 1.